THE GUIDE TO
PAMIR

Theory and Use of Parameterized Adaptive
Multidimensional Integration Routines

THE GUIDE TO
PAMIR

Theory and Use of Parameterized Adaptive Multidimensional Integration Routines

Stephen L. Adler

Institute for Advanced Study, Princeton

World Scientific

NEW JERSEY · LONDON · SINGAPORE · BEIJING · SHANGHAI · HONG KONG · TAIPEI · CHENNAI

Published by

World Scientific Publishing Co. Pte. Ltd.

5 Toh Tuck Link, Singapore 596224

USA office: 27 Warren Street, Suite 401-402, Hackensack, NJ 07601

UK office: 57 Shelton Street, Covent Garden, London WC2H 9HE

British Library Cataloguing-in-Publication Data
A catalogue record for this book is available from the British Library.

THE GUIDE TO PAMIR
Theory and Use of Parameterized Adaptive Multidimensional Integration Routines

ISBN 978-981-4425-03-2
ISBN 978-981-4425-04-9 (pbk)

Printed in Singapore by Mainland Press Pte Ltd.

Dedicated with love to Noah, Tessa, Isabel, and Julian

Contents

Chapter 1

Introduction

PAMIR (Parameterized Adaptive Multi-dimensional Integration Routines) is a suite of Fortran programs for multi-dimensional numerical integration in general dimension p, intended for use by physicists, applied mathematicians, computer scientists, and engineers. The programs are capable of following localized peaks and valleys of the integrand, by subdividing the integration region by a 2^p subdivision at each stage of refinement. An adaptive mesh refinement or "thinning" option allows selection of subregions to be further subdivided, while not further dividing subregions that obey an error criterion. The focus of the PAMIR programs is on integration over hypercubes, simplexes, and hyper-rectangles; large classes of multi-dimensional integrals can be smoothly mapped onto integrals over these base regions.

Our writing the PAMIR programs was motivated by the observation that computer speed has dramatically increased in recent years, while the cost of memory has simultaneously dramatically decreased. Our current laptop speeds, and memories, are characterized by "giga" rather than the "mega" of two decades ago. Moreover, the use of parallel-processing clusters has become commonplace, so with this in mind, every PAMIR routine comes with a parallel MPI (Message-Passing Interface) version. Despite the computational demands of full 2^p subdivision, we have obtained good results from these programs in dimensions up to $p = 7$ on a personal computer, and in dimensions up to $p = 9$ on a cluster. On large clusters with ample running time, integrations over higher dimensions than $p = 9$ are feasible.

The PAMIR programs are available for download on the Internet at the website www.pamir-integrate.com, and are free for large categories of research and U.S. Government users (see Sec. 1.1 for precise details). The purpose of this book is threefold. The first is to provide a manual for use of PAMIR (Chapter 2), and to give specific benchmark and comparison examples (Chapter 3), which serve as illustrations of PAMIR in action. The second is to explain computational integration aspects of the programs, in comparison with integration methods currently in use (Chapter 4). The third is to give a detailed exposition of the algorithms used to construct the PAMIR programs (Chapter 5 and the Appendices). The material given here updates our earlier posting arXiv:1009.4647, and expands on the expository material already on the PAMIR website.

1.1 Copyright, credits, feedback and updates, licensing, and disclaimer

Copyright: All programs are copyright 2010, copyright 2011 by Stephen L. Adler, with the following two exceptions: In cubesubs357.for and cubesubs357m.for (and the real(16) versions of these), the subroutine varsort is a minimal rewriting of the public domain program sort3.f given on John Mahaffy's website. In the tiling programs included in the simplex folders, the subroutine BestLex is the public domain program available on H. D. Knoble's website, with a detailed pedigree given as a comment in the program listing. I am grateful to Prof. Mahaffy and Prof. Knoble for permission to use these subroutines in PAMIR.

Credits: In any publication making use of the PAMIR programs, acknowledgment should be made to use of PAMIR, with a footnote or a parenthesis referring the reader to the PAMIR website www.pamir-integrate.com.

Feedback and updates: We have tested the programs extensively, but would like to hear about any problems users may find. Please send details by email to the author at adler@ias.edu. We will study the problem and make corrections to the program archive as needed. Updates will be noted in a folder "updates", which will be added to the program archive when the first update is made.

Licensing: The PAMIR programs are available free for research use by researchers at not-for-profit colleges and universities, federally funded research facilities, Department of Energy and National Science Foundation Laboratories, User Facilities, and Supercomputer Centers, and for other U.S. Government agencies and facilities, to be used solely for internal, non-commercial research. For downloading the programs onto individual computers, the individual equipment owner should register. For downloading the programs onto institutional networks, the relevant computer manager should register. This license is non-exclusive and non-transferable and does NOT permit redistribution of the Fortran programs, or their translations into other computer programming languages, by electronic or other means; all users are required to register and obtain PAMIR from the website www.pamir-integrate.com. This free use license is not a GNU General Public License.

All other use requires payment of a licensing fee, through the Princeton University Office of Technology Licensing (OTL), after a 30 day free trial period, at the commercial rate. These fees apply to downloading the programs for internal on-site use; licensing for commercial redistribution of the programs must be negotiated on a case-by-case basis with OTL. Contact information for the OTL office and a fee schedule are given on the PAMIR website.

Those interested in obtaining a commercial license must send an email request to the AUTHOR, Dr. Stephen Adler, at adler@ias.edu. The email request should include your full corporate contact information, including company email contact, and provide acknowledgement that you/your company will not use the PAMIR software after the 30 day trial period without execution of the license agreement and payment of the license fee. Once receipt of this email request has been acknowledged, you may go to the registration link to register and download the source code. After such 30 day period, no further use is permitted unless a license agreement with the Princeton University Office of Technology Licensing has been executed and the licensing fee specified in the fee schedule has been paid in full.

Disclaimer of warranty: The PAMIR Programs are provided "as is" and any and all warranties, conditions, representations (express or implied, oral or written), are hereby disclaimed including without limitation any and all implied warranties of quality, performance, merchantability or fitness for a particular purpose. The AUTHOR, Princeton University, and the Institute for Advanced Study do not warrant that the functions contained in the program will meet USER's requirements or that operation will be uninterrupted or error free. The entire risk as to the quality and performance of the PAMIR Programs is with the USER.

Indemnification: USER will indemnify, defend, and hold harmless the AUTHOR, Princeton University, and the Institute for Advanced Study, their trustees, officers, faculty, employees, members, visitors, students, and agents, from and against any and all actions, suits, claims, proceedings, demands, prosecutions, liabilities, costs, expenses, damages, deficiencies, losses or obligations (including attorneys' fees and costs) based on, arising out of, or relating to USER's performance under the Licensing Agreement, including, without limitation USER's use of the PAMIR Programs and any asserted violations of the Export Control Laws by USER.

Application of the PAMIR Programs to problems, the incorrect solution of which may result in personal injury or loss of property, is at your own risk. USER acknowledges and agrees that the AUTHOR, the Institute for Advanced Study, and Princeton University shall not be liable for any direct, indirect, special, incidental, or punitive damages, including lost profits resulting from use of the PAMIR Programs.

1.2   Program language and systems considerations

The PAMIR programs use Fortran 77 conventions for comments and do loops, but also use Fortran 90 constructs such as allocatable memory and array constructors. The total length of the programs in all folders is about 1.1 MB. The real(8) (double precision) serial versions compile and run on the Institute for Advanced Study Linux system with the Intel (compile command: ifort), Portland Group (compile command: pgf90) and GNU version 4.1.2 (compile command: gfortran) Fortran compilers. The real(16) versions compile and run with the Intel compiler, and the MPI versions compile and run with the Intel MPI compiler front end (compile command: mpif90) as well as in cluster operation. (The Linux systems on which the MPI programs were tested have Open MPI installed.) On the author's various personal computers, the real(8) programs compile and run with the Digital Visual Fortran and Intel Visual Fortran packages for Microsoft Windows. (Please note: some of the descriptive names used in this paragraph are registered Trademarks of the vendor companies.)

1.3 Acknowledgements

I wish to thank the Institute for Advanced Study (IAS) School of Natural Sciences computing staff, Prentice Bisbal, Kathleen Cooper, Christopher McCafferty, and James Stephens, for their helpful support and advice throughout this project. I particularly want to acknowledge helpful conversations with Prentice Bisbal about Fortran language features, and to thank my academic assistant Susan Higgins for drawing the figures. I wish to thank the IAS manager of computing Jeffrey Berliner for initiating work on the website, and David Hernandez of the IAS computing administration for the detailed work of setting it up. Momota Ganguli and Judy Wilson-Smith of the IAS Mathematics-Natural Sciences library helped me locate needed references, Roy Gordon suggested additional references, and the anonymous readers' reports gathered by Simon Capelin of Cambridge University Press gave much useful advice, which figured in revising and expanding my initial arXiv posting into this book. I wish to thank my wife, Sarah Brett-Smith, for her patience during the writing and debugging of the programs, and for advice about the ordering of material in this book. I am grateful to John Ritter of the Princeton University Office of Technology Licensing (OTL) for initially suggesting that I distribute the PAMIR programs via a web download, and for helpful advice on implementing this, and to William Gowen of OTL for help in the final stages of website editing. I also wish to thank my editor at World Scientific, Lakshmi Narayanan, and the World Scientific production specialist Rajesh Babu, for their help in the final production process.

I wish to thank Herman D. Knoble for permission to use his public domain program BestLex in the hypercube tiling routines, and John Mahaffey for permission to use a minimal rewriting of his public domain program sort3.f in the subroutines for the cube357 programs. I also wish to thank Bill Press and Saul Teukolsky for emails that helped me implement a function call counter in the programs VEGAS and MISER from their book *Numerical Recipes in Fortran 77*, and to thank the authors of CUBPACK for making an internet download of their programs available.

Program listings are typeset through use of the Listings package of Carsten Heinz and Brooks Moses (`http://mirror.unl.edu/ctan/macros/latex/contrib/listings/listings.pdf`) using the "language=[77]Fortran" option.

Work on the PAMIR programs, and their documentation, was supported in part by the U.S. Department of Energy under grant DE-FG02-90ER40542. I also wish to acknowledge the hospitality of the Aspen Center for Physics during the summers of 2009 through 2012, and the Center's support by the National Science Foundation under Grant No. PHYS-1066293.

Chapter 2

Using PAMIR

In this chapter we survey the basic information needed for a user to get started working with PAMIR. We begin with a description of the integration base regions assumed in the various programs, and the corresponding normalization conventions for integrals. We then give a sketch of how the algorithms work, followed by a list of the contents of the programs in the various folders. We next give a more detailed description of the program inputs and outputs, and show how the outputs are used in making memory and time estimates. After this, we give a discussion of "false positives" encountered when using adaptive thinning, and their avoidance. Finally, we describe features of the subroutine packages that are not accessed by the user in normal operation.

2.1 Base regions for the PAMIR programs, and normalization conventions

2.1.1 *Base region definitions*

The PAMIR program base regions (apart from the hyper-rectangle case listed last) are p-dimensional generalizations of the one-dimensional interval $(0, 1)$, or of the reflection symmetric one-dimensional interval $(-1, 1)$. The comment line headers in each main program state the choice of base region used in that program.

Throughout the book, we use boldface to denote p-vectors, and enumerate the $p + 1$ vertices of a simplex using the letters a, b, c, d

which range from 0 to p, as in $\mathbf{x}_a$, $a = 0, \ldots, p$. The p components of a p-vector are set in italic, not boldface, and are enumerated with the letter i which ranges from 1 to p, as in x_i, $i = 1, \ldots, p$. So a general p-vector $\mathbf{x}$ can be written in terms of components as $\mathbf{x} = (x_1, x_2, \ldots, x_p)$.

(1) *Unit or side 1 hypercube*

The side 1 hypercube base region is defined by the p-fold direct product of intervals $(0, 1) \otimes (0, 1) \otimes \cdots \otimes (0, 1)$, and has volume 1. The integral of a function over the side 1 hypercube has the form

$$
\int_{\text{side 1 hypercube}} dx_1 \cdots dx_p f(x_1, \ldots, x_p)
$$

$$
= \int_0^1 dx_1 \int_0^1 dx_2 \int_0^1 dx_3 \cdots \int_0^1 dx_{p-1} \int_0^1 dx_p f(x_1, \ldots, x_p).
$$

$$(2.1)$$

(2) *Half-side 1 hypercube*

The half-side 1 hypercube base region is defined by the p-fold direct product of intervals $(-1, 1) \otimes (-1, 1) \otimes \cdots \otimes (-1, 1)$, and has volume 2^p. The integral of a function over the half-side 1 hypercube has the form

$$
\int_{\text{half-side 1 hypercube}} dx_1 \cdots dx_p f(x_1, \ldots, x_p)
$$

$$
= \int_{-1}^1 dx_1 \int_{-1}^1 dx_2 \int_{-1}^1 dx_3 \cdots \int_{-1}^1 dx_{p-1} \int_{-1}^1 dx_p f(x_1, \ldots, x_p).
$$

$$(2.2)$$

(3) *Unit standard simplex*

The standard simplex base region has the $p + 1$ vertices

$$
\begin{aligned}
\mathbf{x}_0 &= (0, 0, 0, \ldots, 0), \\
\mathbf{x}_1 &= (1, 0, 0, \ldots, 0), \\
\mathbf{x}_2 &= (0, 1, 0, \ldots, 0), \\
\mathbf{x}_3 &= (0, 0, 1, 0, \ldots, 0), \\
&\quad \cdots\cdots\cdots\cdots \\
\mathbf{x}_{p-1} &= (0, 0, 0, \ldots, 0, 1, 0), \\
\mathbf{x}_p &= (0, 0, 0, \ldots, 0, 0, 1),
\end{aligned}
$$

$$(2.3)$$

and has volume $1/p!$. In terms of a general point $\mathbf{x}$ with components x_i, it is bounded by axis-parallel hyperplanes $x_i = 0$, $i = 1, \ldots, p$ and the diagonal hyperplane $1 = x_1 + x_2 + \cdots + x_p$. Thus, the integral of a function $f(x_1, \ldots, x_p)$ over the standard simplex can be written as a multiple integral in the form

$$
\int_{\text{standard simplex}} dx_1 \cdots dx_p \, f(x_1, \ldots, x_p)
$$

$$
= \int_0^1 dx_1 \int_0^{1-x_1} dx_2 \int_0^{1-x_1-x_2} dx_3 \cdots \int_0^{1-x_1-x_2-\cdots-x_{p-2}} dx_{p-1}
$$

$$
\times \int_0^{1-x_1-x_2-\cdots-x_{p-1}} dx_p \, f(x_1, \ldots, x_p) \,. \tag{2.4}
$$

(4) *Unit Kuhn simplex*

The Kuhn (1960) simplex base region has the $p + 1$ vertices

$$
\begin{aligned}
\mathbf{x}_0 &= (0, 0, 0, \ldots, 0) \,, \\
\mathbf{x}_1 &= (1, 0, 0, \ldots, 0) \,, \\
\mathbf{x}_2 &= (1, 1, 0, \ldots, 0) \,, \\
\mathbf{x}_3 &= (1, 1, 1, 0, \ldots, 0) \,, \\
&\quad\cdots\cdots\cdots\cdots \\
\mathbf{x}_{p-1} &= (1, 1, 1, \ldots, 1, 1, 0) \,, \\
\mathbf{x}_p &= (1, 1, 1, \ldots, 1, 1, 1) \,,
\end{aligned} \tag{2.5}
$$

and has volume $1/p!$. These vertices define a simplex in which $1 \geq x_1 \geq x_2 \geq x_3 \cdots \geq x_{p-1} \geq x_p$. The integral of a function $f(x_1, \ldots, x_p)$ over a unit Kuhn simplex can be written as a multiple integral in the form

$$
\int_{\text{unit Kuhn simplex}} dx_1 \cdots dx_p f(x_1, \ldots, x_p) = \int_0^1 dx_1 \int_0^{x_1} dx_2
$$

$$
\times \int_0^{x_2} dx_3 \cdots \int_0^{x_{p-2}} dx_{p-1} \int_0^{x_{p-1}} dx_p f(x_1, \ldots, x_p) \,. \tag{2.6}
$$

(5) *Hyper-rectangle*

The hyper-rectangle base region is defined by the p-fold direct product of intervals $(a_1, b_1) \otimes (a_2, b_2) \otimes \cdots \otimes (a_p, b_p)$, and has volume $\prod_{i=1}^{p}(b_i - a_i)$. The integral of a function over the hyper-rectangle has the form

$$
\int_{\text{hyper-rectangle}} dx_1 \cdots dx_p f(x_1, \ldots, x_p) = \int_{a_1}^{b_1} dx_1 \int_{a_2}^{b_2} dx_2
$$
$$
\times \int_{a_3}^{b_3} dx_3 \cdots \int_{a_{p-1}}^{b_{p-1}} dx_{p-1} \int_{a_p}^{b_p} dx_p f(x_1, \ldots, x_p). \qquad (2.7)
$$

The programs using a hyper-rectangular base region proceed by mapping it into a half-side 1 hypercube by means of the change of variable, for all i,

$$
y_i = \frac{1}{2}(b_i + a_i) + x_i \frac{1}{2}(b_i - a_i), \qquad (2.8)
$$

which maps the interval $-1 \leq x_i \leq 1$ onto the interval $a_i \leq y_i \leq b_i$.

In two dimensions, the unit standard simplex, side 1 hypercube, and half-side 1 hypercube are illustrated in Fig. 2.1, and the unit Kuhn simplex is illustrated in Fig. 2.2.

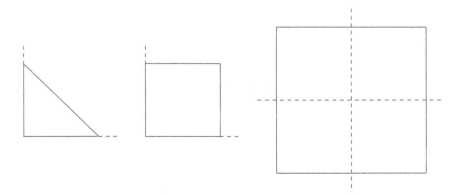

Fig. 2.1 From left to right, the unit standard simplex, the unit or side 1 hypercube, and the half-side 1 hypercube, in two dimensions.

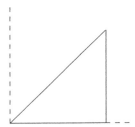

Fig. 2.2   The unit Kuhn simplex in two dimensions.

2.1.2   *Normalization conventions*

For the base regions: side 1 hypercube, half-side 1 hypercube, unit standard simplex, and unit Kuhn simplex, the programs compute the integral normalized by the base region volume,

$$\frac{1}{V_{\text{base region}}} \int_{\text{base region}} dx_1 \cdots dx_p f(x_1, \ldots, x_p), \qquad (2.9)$$

so that for $f \equiv 1$ the normalized integral is 1.

To recapitulate, the base regions volumes are:

$$\begin{aligned} V_{\text{side 1 hypercube}} &= 1, \\ V_{\text{half-side 1 hypercube}} &= 2^p, \\ V_{\text{standard simplex}} &= 1/p!, \\ V_{\text{Kuhn simplex}} &= 1/p!. \end{aligned} \qquad (2.10)$$

For the hyper-rectangular base region, the programs compute the *un-normalized* integral of Eq. (2.7).

2.1.3   *Base region applications*

The choice of base regions is motivated by a number of applications.

(1) An important physics application of the standard simplex is the Feynman–Schwinger formula for combining perturbation theory denominators,

$$\frac{1}{D_0 D_1 \cdots D_p} = p! \int_{\text{standard simplex}} dx_1 \cdots dx_p$$

$$\times \frac{1}{[(1 - x_1 - x_2 - \cdots - x_p)D_0 + x_1 D_1 + \cdots + x_p D_p]^{p+1}}, \quad (2.11)$$

which is proved in Appendix A, and is used in Chapter 3 to formulate test examples.

(2) The integral over a general simplex can be expressed simply as an integral over a standard simplex, using a mapping given in Appendix D (and applied in Appendix B to deriving the simplex generating function).

(3) The integral of a function over a standard simplex can be rewritten as an integral over the unit hypercube by using mappings given in Appendix D.

(4) An integral of the function $f(x_1, \ldots, x_p)$ over the unit hypercube,

$$\int_0^1 dx_1 \int_0^1 dx_2 \cdots \int_0^1 dx_{p-1} \int_0^1 dx_p f(x_1, \ldots, x_p), \quad (2.12)$$

can be rewritten as an integral over $p!$ Kuhn simplexes. To do this, one partitions the hypercube into $p!$ regions, each congruent to the unit Kuhn simplex, by the requirement that in the region corresponding to the permutation P of the coordinate labels $1, \ldots, p$, the coordinates are ordered according to $x_{P(1)} \geq x_{P(2)} \geq x_{P(3)} \cdots \geq x_{P(p-1)} \geq x_{P(p)}$. This partitioning or tiling is illustrated for a square in Fig. 2.3, and for a cube in three dimensions in Fig. 2.1 of Plaza (2007). Hence the integral of f over the unit hypercube is equal to the integral of the *symmetrized* function computed from f, integrated over the unit Kuhn simplex,

$$\int_0^1 dx_1 \int_0^1 dx_2 \cdots \int_0^1 dx_{p-1} \int_0^1 dx_p f(x_1, \ldots, x_p)$$

$$= \int_{\text{unit Kuhn simplex}} dx_1 \cdots dx_p \sum_{p! \text{ permutations } P} f(x_{P(1)}, \ldots, x_{P(p)}).$$

$$(2.13)$$

We will use this equivalence as an alternative method for integration over a unit hypercube.

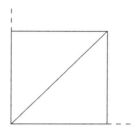

Fig. 2.3   Kuhn simplex tiling of a unit hypercube in two dimensions.

2.1.4   *Remarks on strategy*

- The reason for focusing the programs on base regions that generalize the one-dimensional intervals $(0,1)$ or $(-1,1)$ is that when these regions are successively halved in size, the boundaries are always characterized by fractions of the form $1/2^\ell$. This allows use of integer(2) arithmetic in storing region boundaries, resulting in a significant reduction in memory use.
- Some simple geometric facts are important in setting a strategy for handling hypercubes by analogy with the methods used for simplexes. For a p-dimensional simplex, the number of vertices is $p+1$ which grows linearly with p, permitting one to formulate efficient higher order integration formulas for simplexes that use the vertex coordinates as reference points. For a p-dimensional hypercube, the number of m-dimensional hypercubes on the boundary (see, e.g., the Wikipedia article on hypercubes) is $2^{p-m}p!/(m!(p-m)!)$, and so the number of vertices ($m=0$) is 2^p which grows exponentially with p. Thus, if one constructs hypercube integration formulas in terms of their vertex coordinates, an exponentially growing number of function calls is required for large p. However, for the maximal boundary hypercube, with $m = p-1$, the number given by the above formula is just $2p$, which again has linear growth. (For example, a square has $2 \times 2 = 4$ lines as sides, and a cube has $2 \times 3 = 6$ squares as faces.) So our direct integration method for hypercubes will use geometric features of the maximal boundary hypercubes for

indexing, subdivision, and integration, closely following the methods that we develop for simplexes.

2.2   Sketch of the algorithms

2.2.1   *Folders in pamir_archive*

Downloading PAMIR from the website www.pamir-integrate.com will give the user a folder (or directory) pamir_archive that contains all the programs. The programs are organized into folders cube13, cube357, cube357_16, cube579, cube579_16, simplex123, simplex4, simplex579, simplex579_16, loop_examples, readme, and constant_jacobian_map. The numbers in each folder name indicate the orders of integration routines that are included, for example, in cube579, hypercube integration routines of order 5, 7, and 9 are implemented. Similarly, in simplex579 simplex integration routines of order 5, 7, and 9 are implemented, for both standard and Kuhn simplexes. The subscript _16 indicates that these folders contain quadruple precision (real(16)) versions of the default double precision programs. The folder cube357 contains "hybrid" programs that have an option for subdividing at each stage only some number $p_1 < p$ of the dimensions of a hyper-rectangle, chosen according to various criteria, and of choosing whether to use Monte Carlo or higher order integration to evaluate each subregion. The folder loop_examples contains examples of looping over parameter values in the integrand. The folder readme contains instructions for changing the preset sampling parameters for higher order integration, and need not be accessed if the default parameters are used. Finally, the folder constant_jacobian_map contains main programs implementing a constant jacobian transformation from simplex to hypercube and vice versa; these did not perform well in our tests and so have been removed from the cube and simplex folders and grouped separately for users who may wish to further test them.

The simplex folders also contain programs for integrating over hypercubes by tiling them with Kuhn simplexes, which are useful for dimensions up to to $p = 7$ (beyond this, the combinatorial factor

of $p!$ is too costly, and the direct cube programs should be used.)
Conversely, the cube folders (except for cube13) contain programs for
mapping simplexes into hypercubes by the widely used polynomial
map given in the classic book Stroud (1971), which performed well
in our tests. A detailed discussion of mappings between base regions
is given in Appendix D.

The higher order integration routines use sampling of the integrand at points which are adjustable parameters, input through
the array constructors in the setparam subroutines, with integration
weights which are computed internally; no tables of fixed nodes and
weights appear in the programs. To get a different sampling, one
can change the parameter values within specified limits, as described
in the comment lines for the setparam subroutines and the folder
readme.

2.2.2 *Entering the function and the spatial dimension*

As noted in Chapter 1, the programs are all given as Fortran source
code. The function to be integrated is entered in a subroutine
$fcn(ip, x)$ [or in certain indicated cases, $fcn1(ip, x)$] in the main
program package, with ip the spatial dimension and $x(1{:}ip)$ the integration variable. As set up, each main program package contains
a one line test example in fcn [or $fcn1$]; users can either remove
this example and type or paste in their own function, or remove
the entire subroutine fcn [or $fcn1$] and link to their own $fcn(ip, x)$
[or $fcn1(ip, x)$], written with the same type declarations as in the
example.

The spatial dimension ip and various options for evaluating the
integral are entered in the main program proper, which appears at
the end of each main program package. Both the header of the main
program package, and various programs in it, contain comment lines
giving instructions. The choice of sampling parameters for higher
order integration, and of the comparison function used for thinning,
are also made in the main program package. Under normal use the
subroutine packages do not have to be accessed by the user. If users
wish to modify the subroutines in any way, we strongly recommend

doing a few test integrals before and after the modification. Since users will be editing the main program package, we recommend keeping an original copy from the download on hand, in case the copy being edited is spoiled.

2.2.3   *Basic algorithm module*

The basic algorithm starts from a base region, which acts as the initial level subregion, which is either a standard simplex, a Kuhn simplex, or a half-side 1 hypercube. (Side 1 hypercubes are treated by tiling with Kuhn simplexes, and hyper-rectangles by transformation to half-side 1 hypercubes, as discussed in Secs. 2.1.1, 2.1.3, and Appendix D.) It then proceeds recursively through higher levels of subdivision, by evaluating the integral using an integration method of order specified by the user with two different parameter choices, giving two estimates of the integral over the subregion divided by the subregion volume, which we denote by I_a(subregion) and I_b(subregion). (Dividing out the volume is convenient because of the $1/V$ factor appearing on the left-hand side of Eqs. (5.35) and (5.68) below.) If the level number exceeds a user-specified value *ithinlev* which determines when thinning begins, then a thinning condition is applied. When the user-specified thinning function parameter *ithinfun* is given the value 1, the thinning condition used is

$$|I_a(\text{subregion}) - I_b(\text{subregion})| < \epsilon \,, \tag{2.14}$$

with ϵ an error measure specified by the user. (Further thinning options will be discussed shortly.) If this condition is met, the results are retained as contributions to the I_a and I_b estimates of the integral divided by the base region volume, and the subregion is not further subdivided. If this condition is not met, then the subregion is subdivided into 2^p subregions, and the process is repeated. The process terminates when either the thinning condition is met for all subregions, or a limit to the number of levels of subdivision set by the user is reached. In the latter case, the contributions of the remaining subregions that have not satisfied the thinning condition are added to the I_a and I_b totals, as well as to the sum of the absolute values of the local subregion errors.

2.2.4 *Three versions of the algorithm*

Three versions of the basic algorithm are presented in each of the folders of programs. Main programs that do not end in "m.for" or "r.for" implement the basic algorithm module, in which the algorithm subdivides until all subregions obey the thinning condition, or until a preset limit on the level of subdivisions is reached, which is dictated by the available memory. Typically, for simple integrands and moderate dimension p, this happens rather quickly, in other words, the algorithm has saturated capabilities of the machine memory, but not of the machine speed.

In main programs ending in "r.for", the basic algorithm module is "recirculated" by keeping, at a level limit set by the user which is chosen to avoid exceeding machine memory capabilities, all the subregions that do not obey the thinning condition after processing by the basic module. These subregions are then treated one at a time by the same basic algorithm module, up to a second level limit again set by the user. This can take hours or days for high accuracy, high p computations, with a practical limit set by the speed capabilities of the machine.

Finally, main programs ending in "m.for" are a MPI (Message-Passing Interface) adaptation of the "r.for" programs, in which the residual subregions are distributed to the available parallel processes by a round-robin distribution, analogous to a dealer dealing cards to a group of people seated around a table; this method of distribution helps equalize load distribution. The distributed subregions are then treated sequentially by the parallel processes, again using the basic algorithm module. This speeds up the computation by a factor roughly equal to the number of processes available minus one.

2.2.5 *Default double precision and real(16) option*

All routines are coded in double precision $\big(\text{real}(8)\big)$, but since the ninth order integration formulas involve large numbers in computing coefficients, double precision computation is not enough to give double precision accuracy results, so for the fifth, seventh, and ninth order routines in both the simplex and direct hypercube cases, we also give a quadruple precision $(\text{real}(16))$ version of the programs. However, on the Institute for Advanced Study machines with quadruple precision capability, we found that execution speed for quadruple precision was much slower (often by a factor of 20) than the double precision versions, so we did not run benchmark examples in quadruple precision. We also note that in the MPI programs, the real(16) versions only give double precision results, because the MPI_Send and MPI_Recv statements only permit real(8) variables to be sent and received.

2.2.6   *Linking main programs to subroutines*

In addition to the main programs, each folder contains two subroutine packages, each of which has "subs" as part of the program name. (The one exception to this is the folder constant_jacobian_map, which contains only main programs.) To obtain an executable program, **one** main program should be compiled and linked to **one** of the two subroutine packages. The serial main program packages with names ending in "N.for" and "Nr.for" should be linked to the subroutine package ending in "N.for", with "N" the one to three digit integration order label. The MPI parallel main program packages with names ending in "Nm.for" should be linked to the corresponding subroutine package ending in "Nm.for". (Instructions for linking the main programs in the folder constant_jacobian_map are given in Table 2.1.)

2.2.7   *Error estimates and thinning*

At termination of the basic algorithm module, we get the final estimates of the integral divided by the base region volume,

$$
\begin{aligned}
I_a &\simeq \sum_{\text{subregions}} V(\text{subregion}) I_a(\text{subregion}) , \\
I_b &\simeq \sum_{\text{subregions}} V(\text{subregion}) I_b(\text{subregion}) .
\end{aligned}
\tag{2.15}
$$

Here $V(\text{subregion})$ is the subregion volume divided by the base region volume, and since the subregions are a tiling of the initial base region, we have

$$\sum_{\text{subregions}} V(\text{subregion}) = 1 \,. \qquad (2.16)$$

From the difference of I_a and I_b we get an estimate of the error, given by

$$|\text{outdiff}| \equiv |I_a - I_b| \,. \qquad (2.17)$$

We also compute the sum of the absolute values of the local subregion errors,

$$\text{errsum} \equiv \sum_{\text{subregions}} V(\text{subregion})|I_a(\text{subregion}) - I_b(\text{subregion})| \,.$$
$$(2.18)$$

Comparing Eqs. (2.15), (2.17), and (2.18), we see that errsum and $|\text{outdiff}|$ obey the inequality

$$|\text{outdiff}| \leq \text{errsum} \,, \qquad (2.19)$$

with equality holding if $I_a - I_b$ has the same sign in all subregions. If the condition $|I_a(\text{subregion}) - I_b(\text{subregion})| < \epsilon$ is met for all subregions, using Eqs. (2.14), (2.16), and (2.19) we get

$$|\text{outdiff}| \leq \text{errsum} < \epsilon \,. \qquad (2.20)$$

When this holds, to evaluate the integral to a relative error δ, one should choose

$$\epsilon \sim \delta |I_a| \,. \qquad (2.21)$$

However, if it is not feasible to subdivide finely enough to satisfy the condition $|I_a(\text{subregion}) - I_b(\text{subregion})| < \epsilon$ for all subregions, errors are determined by the coarseness of the subregions, and the ϵ of Eq. (2.21) may be unnecessarily small.

Using $|I_a(\text{subregion}) - I_b(\text{subregion})|$ as the basis for a thinning decision is only one possibility of many. More generally, given $A \equiv I_a(\text{subregion})$ and $B \equiv I_b(\text{subregion})$, one can take as the thinning function any function $f(A, B)$ with the properties $f(A, B) \geq 0$ and $f(A, B) = 0$ iff $A = B$, imposing now the thinning condition $f(A, B) < \epsilon$. In the programs, we have included three options, (1) $f(A, B) = |A - B|$ as in the discussion above,

(2) $f(A, B) = |A - B|/|A + B|$, and (3) $f(A, B) = (A - B)^2$. In many cases, and in particular for polynomial integrals, we found their performance (with appropriate ϵ) to be similar, but for the singular integral $\int_0^1 dx \frac{1}{\sqrt{1-x^2}}$ we found choice (3) to perform considerably better than the other two. We do not have examples in which choice (2) is distinctly better.

2.2.8   *User options*

The programs present the user with various options. The error measure ϵ introduced in the preceding subsection is set by a user supplied parameter *eps*. By an appropriate choice of the parameter *ithinlev*, thinning can be delayed, or even suppressed entirely so that all subdivisions take place to the specified subdivision limits. This can give a check that subregions with large contributions, but accidentally small error estimates, have not been harvested prematurely; for further details, see the discussion of false positives in Sec. 2.6 below. Suppressing thinning is also useful when the programs are modified, to give a check that the tiling condition of Eq. (2.16) is obeyed. By a choice of the parameter *ithinfun*, the user can choose which of three preset thinning functions to use, or by modifying the subroutine containing these functions, the user can make another choice of thinning function. For simplex integration, the parameter *isimplex* chooses whether the base region is a standard or Kuhn simplex, and the parameter *isubdivision* allows the user to choose whether to use the recursive or the symmetric subdivision algorithm discussed below in Chapter 5. For all programs, a parameter *iaccuracy* allows the user to choose which of various higher order integration methods to use. The hybrid programs in the folder cube357 contains further options, described in detail in Tables 2.2 and 2.3 in Sec. 2.4.1 below. The main program parameters that set the maximum number of subdivisions are also explained in Sec. 2.4.1.

2.3   Contents of programs in the folders

We give here a list of the program files in the various folders in the download pamir_archive. Each main program file contains a number of subroutines, including the one in which the function to be integrated is input, with the main program proper at the end following the subroutines. Brief comment lines give instructions for use. The subroutine packages contain multiple subroutines, and do not have to be accessed by the user in normal operation.

- The "simplex" programs take as base region the standard simplex of Eq. (2.3) or the Kuhn simplex of Eq. (2.5).
- The "cubetile" programs take as base region the side 1 hypercube of Eq. (2.1), and integrate by tiling it with Kuhn simplexes.
- The "cube" programs take as base region the half-side 1 hypercube of Eq. (2.2).
- The "simplexmap" programs take as base region the standard simplex or Kuhn simplex, and integrate by making a Stroud map to a half-side 1 hypercube.
- The "rectangle" programs take as base region the general hyperrectangle of Eq. (2.7), and integrate by transformation to a half-side 1 hypercube.
- REMINDERS: (i) The numbers after "simplex" or "cube" in the folder names indicate the integration orders that are included. (ii) Except for the "rectangle" programs, integrals are normalized to the base region volume, as discussed in Sec. 2.1.2. (iii) One main program should be linked to one subroutine package: those not ending in m.for are linked to the non-m.for subroutine package, those ending in m.for are linked to the m.for subroutine package, as discussed in Sec. 2.2.6.

2.3.1   *Folder simplex123*

This folder contains:

- Subroutine packages simplexsubs123.for, simplexsubs123m.for.

- Main program files simplexmain123.for, cubetilemain123.for.
- Recirculating main program files simplexmain123r.for, cubetilemain123r.for.
- MPI parallel main program files simplexmain123m.for, cubetilemain123m.for.

2.3.2 *Folder simplex4*

This folder contains:

- Subroutine packages simplexsubs4.for, simplexsubs4m.for.
- Main program files simplexmain4.for, cubetilemain4.for.
- Recirculating main program files simplexmain4r.for, cubetilemain4r.for.
- MPI parallel main program files simplexmain4m.for, cubetilemain4m.for.

2.3.3 *Folder simplex579*

This folder contains:

- Subroutine packages simplexsubs579.for, simplexsubs579m.for.
- Main program files simplexmain579.for, cubetilemain579.for.
- Recirculating main program files simplexmain579r.for, cubetilemain579r.for.
- MPI parallel main program files simplexmain579m.for, cubetilemain579m.for.

2.3.4 *Folder simplex579_16*

This folder contains:

- Subroutine packages simplexsubs579_16.for,

simplexsubs579_16m.for.
- Main program files simplexmain579_16.for, cubetilemain579_16.for.
- Recirculating main program files simplexmain579_16r.for, cubetilemain579_16r.for.
- MPI parallel main program files simplexmain579_16m.for, cubetilemain579_16m.for.

2.3.5 *Folder cube13*

This folder contains:

- Subroutine packages cubesubs13.for, cubesubs13m.for.
- Main program file cubemain13.for.
- Recirculating main program file cubemain13r.for.
- MPI parallel main program file cubemain13m.for.

2.3.6 *Folder cube357*

This folder contains:

- Subroutine packages cubesubs357.for, cubesubs357m.for.
- Main program files cubemain357.for, rectanglemain357.for, simplexmapmain357.for.
- Recirculating main program files cubemain357r.for, rectanglemain357r.for, simplexmapmain357r.for.
- MPI parallel main program files cubemain357m.for, rectanglemain357m.for, simplexmapmain357m.for.

2.3.7 *Folder cube357_16*

This folder contains:

- Subroutine packages cubesubs357_16.for, cubesubs357_16m.for.
- Main program files cubemain357_16.for, rectanglemain357_16.for, simplexmapmain357_16.for.

- Recirculating main program files cubemain357_16r.for, rectanglemain357_16r.for, simplexmapmain357_16r.for.
- MPI parallel main program files cubemain357_16m.for, rectanglemain357_16m.for, simplexmapmain357_16m.for.

2.3.8 *Folder cube579*

This folder contains:

- Subroutine packages cubesubs579.for, cubesubs579m.for.
- Main program files cubemain579.for, simplexmapmain579.for.
- Recirculating main program files cubemain579r.for, simplexmapmain579r.for.
- MPI parallel main program files cubemain579m.for, simplexmapmain579m.for.

2.3.9 *Folder cube579_16*

This folder contains:

- Subroutine packages cubesubs579_16.for, cubesubs579_16m.for.
- Main program files cubemain579_16.for, simplexmapmain579_16.for.
- Recirculating main program files cubemain579_16r.for, simplexmapmain579_16r.for.
- MPI parallel main program files cubemain579_16m.for, simplexmapmain579_16m.for.

2.3.10 *Folder constant_jacobian_map*

- See Table 2.1 for the main programs in this folder, and the subroutine packages to which they must be linked.

Table 2.1 How to link the main programs in constant_jacobian_map to the appropriate subroutine package (.for is omitted from program names).

main program name	link to	in folder
simplexmap2main579	cubesubs579	cube579
simplexmap2main579r	cubesubs579	cube579
simplexmap2main579m	cubesubs579m	cube579
simplexmap2main579_16	cubesubs579_16	cube579_16
simplexmap2main579_16r	cubesubs579_16	cube579_16
simplexmap2main579_16m	cubesubs579_16m	cube579_16
simplexmap2main357	cubesubs357	cube357
simplexmap2main357r	cubesubs357	cube357
simplexmap2main357m	cubesubs357m	cube357
simplexmap2main357_16	cubesubs357_16	cube357_16
simplexmap2main357_16r	cubesubs357_16	cube357_16
simplexmap2main357_16m	cubesubs357_16m	cube357_16
cubemapmain579	simplexsubs579	simplex579
cubemapmain579r	simplexsubs579	simplex579
cubemapmain579m	simplexsubs579m	simplex579
cubemapmain579_16	simplexsubs579_16	simplex579_16
cubemapmain579_16r	simplexsubs579_16	simplex579_16
cubemapmain579_16m	simplexsubs579_16m	simplex579_16
cubemapmain123	simplexsubs123	simplex123
cubemapmain123r	simplexsubs123	simplex123
cubemapmain123m	simplexsubs123m	simplex123

2.3.11 *Folder loop_examples*

This folder contains:

- Main program simplexmain579r.for (must be linked to subroutine package simplexsubs579.for in folder simplex579).
- Main program simplexmain579m.for (must be linked to subroutine package simplexsubs579m.for in folder simplex579).

- Main program simplexmain579m1.for (an alternative MPI main program, using MPI_REDUCE in place of MPI_Send and MPI_Recv; must be linked to subroutine package simplexsubs579m.for in folder simplex579).

2.3.12   *Folder readme*

This folder contains:

- Text file setparam.txt, giving instructions for changing the parameters governing sampling points for higher order integration. It need not be read if one only wishes to use the default parameter values in the main programs.

2.4   Program inputs and outputs

2.4.1   *Inputs*

Inputs of parameters governing the integration process are inserted in the main program between the two lines of stars (******); lines of code outside this region should not be changed. The comment lines in the programs give brief instructions for entering inputs. When a value for any parameter is input that is outside the required range, the program prints "Warning: a parameter you entered in main is outside limits" and does not execute. In the enumeration of inputs that follows, italics are used for emphasis for the input parameter names; the parameter names appear in Roman in the programs and in the tables of this section.

(1) *ithinlev* tells the program when to begin evaluating and testing subregions. When the total level number is greater than *ithinlev*, the program evaluates the subregion integrals, harvesting those that obey the thinning condition of Eq. (2.14), and subdividing those that do not. When the total level number is less than or equal to *ithinlev*, the program subdivides *without* evaluating the subregion integrals. If *ithinlev* is greater than or equal to the level number limit, there is no thinning; the program subdivides

until the level number limit is reached and then evaluates all subregions.

(2) *ithinfun* instructs the program which thinning function option to use. As explained in Sec. 2.2.7, *ithinfun* = 1 corresponds to a thinning function $f(A, B) = |A - B|$, *ithinfun* = 2 to $f(A, B) = |A - B|/|A + B|$, and *ithinfun* = 3 to $f(A, B) = (A - B)^2$. The thinning function is computed in the function subroutine fthin.

(3) *isimplex* (in the simplex programs) selects a standard simplex base region for *isimplex* = 1, and a Kuhn simplex base region for *isimplex* = 2.

(4) *isubdivision* tells the simplex programs whether to use symmetric subdivision (*isubdivision* = 1) or recursive subdivision (*isubdivision* = 2). These subdivision methods are explained in detail in Chapter 5 below. This parameter does not appear in the "cube" main programs in cube13, cube357, cube579, and the corresponding real(16) versions, where there is no choice of subdivision method.

(5) *iaccuracy* instructs the programs to use the integration program of order *iaccuracy*. For example, in the simplex123 programs, to select third order accuracy one sets *iaccuracy* = 3, and in the simplex 579 programs, to select seventh order integration one sets *iaccuracy* = 7.

CAUTION: This parameter does not appear in the main programs in simplex4, which uses only fourth order integration. The programs in simplex4 perform fourth order adaptive integration using a different subdivision strategy from that used in all the other cases. In simplex4 the simplex vertices are used as sampling points, with the side midpoints giving the vertices at the next level of subdivision. In all the other programs, only interior points of the simplex are used for sampling. Hence, the simplex4 programs cannot be used to integrate functions which have integrable singularities at the base simplex boundary, whereas the other programs can be used in this case.

(6) *ip* gives the spatial dimension p of the simplex or hypercube being integrated over, and can take any integer value ≥ 1. Thus, to integrate over a three-dimensional cube one would set $ip = 3$.

(7) *eps* sets the parameter ϵ appearing in the thinning condition of Eq. (2.14). For *ithinfun*=1, this gives an absolute error criterion; to achieve a given level of relative error, one needs a rough estimate of the value of the integral as given by *outa* or *outb*, which can be used to readjust *eps* by use of Eq. (2.21). For nonsingular integrands, the *eps* value when using *ithinfun*=3 should, as a first guess, be taken as the square of the *eps* value that one used for *ithinfun*=1. Note that if *ithinlev* is greater than or equal to the total level number, so that thinning is suppressed, the results are independent of the value given to *eps*.

(8) In all programs other than the "cubetile" hypercube tiling programs and the various "map" programs, the external function is supplied by the user in the subroutine fcn.for. In the "cubetile" programs, fcn.for is the symmetrization program for the external function supplied by the user in the subroutine fcn1.for. In the "map" programs, fcn.for is the mapping program for the external function supplied by the user in the subroutine fcn1.for.

(9) *llim* sets the limit to the number of subdivisions in the programs with names not ending in "r" or "m". It can be any integer between 1 and 15, except in the simplex4 programs, where the range is 1 to 14. In practice, the effective upper limit is set by machine performance, as discussed below in Sec. 2.5.

(10) *llim1* and *llim2* in the programs with names ending in "r" and "m" set the limit to the number of subdivisions in the first and second stages of subdivision, respectively. The maximum value of *llim1* is 14 and of *llim2* is 15, except in the simplex4 programs, where the respective limits are 13 and 14, and also except for recursive subdivision, where the maximum value of *llim2* is one less than the corresponding value for symmetric subdivision. In all cases, the built-in subdivision limits prevent the program from dividing 1 by 2, which would give an integer arithmetic answer of 0 for a point of the integer(2) lattice defining the subregions. (See the discussion of the use of integer(2) arithmetic in Sec. 2.7.) Again, the effective upper limit is set by machine performance.

(11) The parameters for the higher order integration routines are contained in the main program files in subprograms with names

beginning with "setparam". They are given in array constructors, and have been preset to values indicated. These presets can
be changed by the user to give a different sampling of the integrand in the integration subregions, subject to rules which are
summarized in the file "setparam.txt" in the folder "readme".
In brief, (i) the inequalities in the comment statements must
be obeyed, to keep the sampling points inside the subregion;
(ii) the parameters in each array constructor must have non-
degenerate values, so that the equations for integration weights
will be solvable; (iii) the "a" and "b" array constructors should
have different parameter values, since these are used to give the
two different integrand evaluations used in the error estimate;
(iv) in the 357 hybrid folder, some of the parameters are used to
compute finite differences along the axis directions, resulting in
additional restrictions detailed in "setparam.txt".

(12) The hybrid programs in the folder cube357 include the additional options of comparing Monte Carlo evaluation of a subregion to higher order integration evaluation, and of subdividing
only along some number $ip1 \leq ip$ of the axes, chosen according
to various criteria. In addition to the new parameter $ip1$, this
entails further input parameters imc, $iflagopt$, $idiff$, $eps1$, $delta$,
$ihit$, and $iseed$ to be set by the user. The parameters imc and
$iflagopt$ take the values 1, 2, or 3, and together with the real
parameters $eps1 > 0$ and $delta > 0$ control the subdivision logic
as shown in Tables 2.2 and 2.3. The parameter $idiff$ controls the
order $= 2 \times idiff$ of the finite difference var(i) used to govern
subdivision of the axis i; $idiff$ can take the values 1 or 2 for
$iaccuracy = 3$, the values 1, 2, or 3 for $iaccuracy = 5$, and the
values 1, 2, 3, or 4 for $iaccuracy = 7$. The parameter $ihit$ is the
number of Monte Carlo samplings used for each of two estimates
of the subregion integral, and the parameter $iseed$ is the random number seed used in the Monte Carlo option. Some further
remarks on the hybrid programs are:

(a) For $l \leq ithinlev$, there is a full 2^p subdivision. The "r"
and "m" versions require $ithinlev \geq llim1$, and so in the

first stage there is a full 2^p subdivision with no thinning. Thinning with the options shown in Tables 2.2 and 2.3 occurs in the second stage.

(b) Choosing $imc = 1$ or $imc = 2$ with $delta = 0$ is pure Monte Carlo, with one of the higher order algorithms still evaluated but used only to determine which axes to subdivide when $imc = 2$.

(c) Choosing $imc = 0$, $iflagopt = 1$, $ip1 = ip$ reduces to the cube program with full 2^p subdivision.

(d) Because the error estimate for Monte Carlo is based on the difference of two samplings $|I_a - I_b|$, it is finite even when the integral of the function squared $\int f^2$ diverges and thus the variance is infinite.

(e) As an illustration of using the options provided by $ip1$, $iflagopt$, and $ithinlev$, suppose one has a function in 5 dimensions that one knows to be rapidly varying in dimensions 1 and 2, so that a fine mesh is required, but slowly varying in dimensions 3, 4, 5, so that a coarse mesh of 8 subdivisions suffices. Then one can take $ip1 = 2$ and $ithinlev = 3$. With these parameters, the program subdivides by a factor of 2 along all 5 axes (this is what we mean by a 2^5 subdivision) in going three levels from level 1 to level 4, resulting in 8 subdivisions in each of the 5 dimensions. After this, the program only further subdivides dimensions 1 and 2, where a finer mesh is needed, using the subdivision strategy specified by $iflagopt$ according to Table 2.2.

2.4.2 *Outputs*

Program outputs (except in the MPI case) are written to a file "outdat.txt" and also appear on the screen. In the MPI case, outputs are written to the output file specified by the system for a "print" statement. A brief description of output labeling follows (again, italics are used here for emphasis; the output names appear in Roman

Table 2.2   cube357 subdivision options used when the local error criterion is *not* obeyed. Case $imc = 0$: no Monte Carlo, only higher order integration is used.

iflagopt = 1	*iflagopt* = 2	*iflagopt* = 3
Axes $1, \ldots, ip1$ are halved	Axes with the $ip1$ largest integrand variations $var(i)$ are halved	Axes i with $var(i) >$ $eps1 * \max_i[var(i)]$ are halved

Table 2.3   cube357 options used when the local error criterion is *not* obeyed. Case $imc > 0$: Chooses Monte Carlo when $delta * \mathrm{errest}(\text{monte carlo}) < \mathrm{errest}(\text{higher order})$; otherwise chooses higher order with subdivision as in Table 2.2.

$imc = 1$	choosing Monte Carlo followed by full 2^p subdivision
$imc = 2$	choosing Monte Carlo followed by hyper-rectangular subdivision according to Table 2.2

in the programs).

(1) All programs write out the user-set values of *ip*, *llim* (or *llim1* and *llim2*), *eps*, *ithinlev*, *ithinfun*, and *iaccuracy*. The simplex programs also write out user-set values of *isimplex* and *isubdivision*. The programs do not write out the values of the parameters in the array constructors in the subprograms setparam.

(2) In all programs, *outa* and *outb* give two evaluations of the integral normalized as in Sec. 2.1.2, corresponding respectively to the two different samplings of the integrand determined by the "a" and "b" parameters in the array constructors. The output $outav = \frac{1}{2}(outa + outb)$ is the average value of the two samplings, and $|outdiff|$ gives the difference $|outa - outb|$. The size of $|outdiff|$ gives an estimate of the likely error in the answer, but it can be smaller than the true error if the mesh has not been made fine enough. To get reliable error estimates, one should look for stability of the answer as the mesh size (the

number of levels of subdivision) and the error parameter *eps* are changed. With the exception of the "rectangle" programs based on a hyper-rectangle base region, to get the value of the integral without normalization by the base region volume, one must multiply *outa*, *outb*, and *outav* by the base region volume, as described in Sec. 2.1.2.

(3) In all programs, *errsum* gives the sum of the absolute values of the local subregion thinning tests,

$$\text{errsum} \equiv \sum_{\text{subregions}} V(\text{subregion})|f(I_a(\text{subregion}, I_b(\text{subregion})))|,$$

(2.22)

with $f(A, B)$ the thinning function. As explained above, for the choice *ithinfun* = 1 this gives an upper bound for $|outdiff|$, and when $I_a(\text{subregion}) - I_b(\text{subregion})$ has uniform sign over all subregions, $errsum = |outdiff|$. However, when signs are not uniform over subregions, *errsum* for *ithinfun* = 1 can be much larger than $|outdiff|$.

(4) In all programs, *l* gives the level number, *ind* gives the number of subregions carried forward to the next level, *indmax* gives the maximum value of *ind* encountered over the course of the various levels that have been executed, *fcncalls* gives the number of function calls, *t_current* gives the current elapsed time in seconds at the various levels of the first stage, and *t_final* gives the total elapsed execution time in seconds.

(5) In the "recirculating" programs with main program name ending in "r", *t_restart* gives the time at which the second stage is initiated, in which the subregions carried forward from the first stage are subdivided sequentially. During the second stage, the program will indicate when it is approximately 10, 20, ..., 90, 100 percent finished in sequentially processing the subregions carried forward from the first stage, by printing this information to the screen, but not by writing it to file. (Since these numbers are computed by integer division, the actual numbers may be $9, 19, \ldots$ or other similar strings, depending on the residue modulo ten of the number of regions carried forward.) In interactive mode, this permits one to gauge how long the calculation

will take to finish; if it looks like the calculation will take longer than one wishes to wait, one can stop execution and restart with different, more tractable, parameter values. The final statistics include *indcount*, which gives a sum of the *ind* values at each level of the second stage, and which indicates the *ind* value that would be encountered if the second stage subdivisions were carried out in the first stage by using a larger *llim*1 value. Because of the staging strategy, the maximum *ind* value that is encountered in the "r" programs is the much smaller number *indmax*.

(6) In the MPI programs with main program name ending in "m", *t_restart* is the time at the end of the first stage when the subregions carried forward are distributed to multiple processes. The total number of distributed subregions is given by the final first stage value of *ind*, and the final first stage value of *fcncalls* gives the number of function calls up to this point. If *t_restart* does not appear in the output, the program has completed execution before entering the second stage. Since MPI programs are typically run in batch mode, no intermediate statistics are output during the second stage. The final statistics include *indmaxprocess*, which is the maximum of the final *indmax* over all of the second stage processes.

2.5 Function calls, timing estimates, and memory estimates

2.5.1 *Function call counting*

Running time of the PAMIR programs is largely dependent on the number of function calls. Each time a higher order integration subroutine is called, the number of function calls is a number C given in Table 2.4 for the simplex integration programs, and Table 2.5 for the hypercube integration programs. These tables were obtained by running the programs to integrate the function $f = 1$, in which case the programs exit without subdividing the base region, giving the desired function call count for two samplings of the integral at the indicated order of accuracy, as well as unity as the output integral (since the programs compute the integral over the base region, divided by the base region volume). These tables can also be read off from the analytic formulas for the number of function calls given in Table 2.6. For the Monte Carlo option of the cube357 programs, the corresponding number C is two times *ihit*.

The total number of function calls *fcncalls* in running the program in p dimensions is the product of C with a sum S representing the total number of subregions summed over. Since the program subdivides without evaluating integrals for $l < lmin = ithinlev + 1$, the subregion count through the final level of the one or two stage programs is a sum of terms

$$S = \sum_{l=lmin}^{lmax} ind_{l-1}, \quad lmin < lmax,$$

$$(2.23)$$

$$S = ind_{lmax-1}, \quad lmin \geq lmax,$$

where ind_{l-1} is the number of subregions carried forward from level $l-1$ to level l (with $ind_0 = 1$ because the first level always receives one subregion). In applying this formula to the two stage "r" program, $ind_{llim1+l}$ is given by $indcount_l$ for the corresponding second stage level l. The second line of this equation reflects the fact that the final level of each program always evaluates all subregions carried forward to it. The upper limit $lmax$ in this sum is given by $lmax = llim$ for the one stage programs, and $lmax = llim1 + llim2$ for the two stage programs. When thinning is effective, the values of ind_{l-1} are not known *a priori*, and moreover, the intermediate statistic $indcount_l$ is not printed out in the second stage of the "m" programs. But in all versions, to get an upper limit S_{max} on the number of function calls, we use the fact that when there is no thinning of subregions, ind_{l-1} is given by

$$ind_{l-1} = 2^{p(l-1)}.$$

$$(2.24)$$

Substituting this into Eq. (2.23) reduces S_{max} to the sum of a geometric series,

$$S_{max} = \sum_{l=lmin}^{lmax} 2^{p\,(l-1)} = \frac{2^{p\,lmax} - 2^{p\,(lmin-1)}}{2^p - 1}, \quad lmin < lmax\,, \tag{2.25}$$

$$S_{max} = 2^{p\,(lmax-1)}, \quad lmin \ge lmax\,.$$

When thinning is employed, the ratio S/S_{max}, or equivalently $fcncalls/(CS_{max})$, is a measure of how effective thinning has been in reducing running time. Defining K by $K = lmax - 1$, that is,

$$K = llim - 1 \qquad \text{one stage (non} - \text{``r'', ``m'') version}\,,$$
$$K = llim1 + llim2 - 1 \quad \text{two stage (``r'' or ``m'') version} \tag{2.26}$$

we note that when $2^p \gg 1$, the formula of Eq. (2.25) becomes

$$S \simeq 2^{Kp}\,, \tag{2.27}$$

that is, it is dominated by the last term in the series. So as one would expect, increasing the total number of levels of subdivision by 1 multiplies the number of function calls by approximately a factor of 2^p.

As a very rough rule of thumb, in using integration routines of order $n = 2t + 1$ in dimension p, one should avoid high order routines with $t > p$. This is true both because the higher order routines have redundant function calls for low dimension p (see the discussion in Sec. 5.6.6), and because the extra computation involved in using a high order routine is justified only when the 2^p scaling in the number of subregions, as the program subdivides from level to level, becomes large enough. However, this is only a very general criterion, since the optimum choice of integration routine order will depend on the nature of the function being integrated. Moreover, in dimension $p = 1$ the programs are so fast on current computers that use of the fifth or seventh order integration routines gives good results.

2.5.2 *Timing estimates*

Using the results of the preceding section, one can make final running time estimates from the intermediate outputs of the three versions

Table 2.4 Function calls for generating two samplings by the higher order integration subprograms, for simplex integration of order n in dimension p.

n $p \rightarrow$	1	2	3	4	5	6	7	8	9
1	3	4	5	6	7	8	9	10	11
2	5	7	9	11	13	15	17	19	21
3	7	10	13	16	19	22	25	28	31
4	10	16	23	31	40	50	61	73	86
5	20	31	43	56	70	85	101	118	136
7	37	71	117	176	249	337	441	562	701
9	74	168	316	531	827	1219	1723	2356	3136

Table 2.5 Function calls for generating two samplings by the higher order integration subprograms, for hypercube integration of order n in dimension p.

n $p \rightarrow$	1	2	3	4	5	6	7	8	9
1	3	5	7	9	11	13	15	17	19
3	5	9	13	17	21	25	29	33	37
5	12	27	46	69	96	127	162	201	244
7	21	69	153	281	461	701	1009	1393	1861
9	48	192	501	1059	1966	3338	5307	8021	11644

of the programs. Since timing values can vary by one or two tenths of a second between identical runs, estimates of total running time become reproducible only when one has proceeded to the point where several seconds have elapsed. The timing estimates given here are upper bounds, both because they do not take thinning into account, and because the final level does not subdivide, and so runs faster than one would infer from extrapolations based on the earlier levels, which do subdivide.

(1) For the single stage (non-"r","m") programs, from the current running time $t\_current$ at the output of any level l, the maximum running time in the absence of thinning to reach a level

Table 2.6 Function calls for generating two samplings by the higher order integration subprograms. For cubeintI programs $x = 2p$, and for simplexintI programs $x = p + 1$, where "I" is the integration order label.

subroutine	function calls
cubeint1	$x + 1$
cubeint3	$2x + 1$
cubeint5	$x(x - 1)/2 + 5x + 1$
cubeint7	$x(x - 1)(x - 2)/6 + 3x(x - 1)$
	$+7x + 1$
cubeint9	$x(x - 1)(x - 2)(x - 3)/24 + 7x(x - 1)(x - 2)/6$
	$+17x(x - 1)/2 + 15x + 1$
simplexint1	$x + 1$
simplexint2	$2x + 1$
simplexint3	$3x + 1$
simplexint4	$x(x - 1)/2 + 4x + 1$
simplexint5	$x(x - 1)/2 + 9x + 1$
simplexint7	$x(x - 1)(x - 2)/6 + 5x(x - 1)$
	$+13x + 1$
simplexint9	$x(x - 1)(x - 2)(x - 3)/24 + 11x(x - 1)(x - 2)/6$
	$+31x(x - 1)/2 + 21x + 1$

limit $llim > l$ is the product $t\_current$ times $2^{p\,(llim-l)}$.

(2) For the recirculating or "r" programs, one can estimate the total running time, in the absence of second stage thinning, by multiplying the time of starting the second stage $t\_restart$, by $2^{p\,llim2}$ to take into account the second stage subdivisions.

(3) For the MPI or "m" programs, one can estimate the total second stage running time in the absence of second stage thinning, by multiplying $t\_restart$ by $2^{p\,llim2}$ to take into account the second stage subdivisions, and dividing by $N_{\text{process}} - 1$, with N_{process} the number of processes. (Process 0 serves only as an accumulation register for the output of the remaining $N_{\text{process}} - 1$ processes.) If this estimate is too large, one can stop execution of the batch job and restart with less ambitious parameters.

2.5.3 *Memory estimates*

The most memory intensive part of the programs is the storage of the coordinates defining the subregions at a given level of subdivision. As discussed in Sec. 2.7 below, to conserve memory, these are represented as points of an integer(2) lattice and use allocatable memory. In Table 2.7 we give formulas for the size of the largest integer(2) arrays simultaneously present in the various programs, which give rough lower bounds for the total memory requirements. In principle, these formulas can be used to estimate the largest level number attainable by a computer with a given architecture and memory capacity.

In practice, we found it easiest to determine the memory limit empirically, as follows.

(1) For the single stage (non-"r","m") programs, start with a low value of $llim$, and then to improve the accuracy, increase it until you get a diagnostic saying memory has been exceeded; the value of $llim$ one less than this is the maximum value $llim = LMAX$ that does not exceed memory. Since the final level of the single stage program $l = llim$ does not further subdivide, the limit $LMAX$ is associated with the number of subregions carried forward from level $l - 1$ to the final level.

(2) In using the two-stage "r" and "m" programs, and setting $llim2 = 1$, the maximum value of $llim1$ that will not exceed memory is $llim1 = LMAX - 1$, with $LMAX$ the corresponding maximum determined as above for the single stage program. The fact that the maximum $llim1$ is now 1 less than the maximum $llim$ in the single stage case reflects the fact that the final level of the first stage of the two-stage programs does subdivide, before passing the residual subregions to the second stage. Once $LMAX$ is determined, one can take any value $1 \leq llim2 \leq LMAX$ without exceeding memory. Because of the way we have set up the staging, the numerical output depends only on the sum $llim1 + llim2$, that is, one is free to redistribute the computational effort between the first and second stages.

Table 2.7 Largest integer(2) array storage requirements for the various programs at level ℓ, as a function of the spatial dimension p and the number $ind_{(\ell-1)}$ of subregions carried into that level, and the number ind_ℓ of subregions carried forward to the next level $\ell + 1$. When there is thinning, $ind_\ell \leq 2^p \, ind_{(\ell-1)}$, while when there is no thinning, $ind_\ell = 2^p \, ind_{(\ell-1)}$, giving the second column of the table. This table does not include storage requirements for the smaller arrays, and so gives a rough lower bound on the memory requirements. The real(16) versions of the programs have the same integer(2) storage requirements as the real(8) versions in the table.

program	integer(2) storage locations with thinning	integer(2) storage locations without thinning
cube13	$(p+1)[ind_\ell + 2^p \, ind_{(\ell-1)}]$	$2(p+1) \, ind_\ell$
cube357	$2p[ind_\ell + 2^p \, ind_{(\ell-1)}]$	$4p \, ind_\ell$
cube579	$(p+1)[ind_\ell + 2^p \, ind_{(\ell-1)}]$	$2(p+1) \, ind_\ell$
simplex123	$p(p+1)[ind_\ell + 2^p \, ind_{(\ell-1)}]$	$2p(p+1) \, ind_\ell$
simplex4	$p(p+1)[ind_\ell + 2^p \, ind_{(\ell-1)}]$	$2p(p+1) \, ind_\ell$
simplex579	$p(p+1)[ind_\ell + 2^p \, ind_{(\ell-1)}]$	$2p(p+1) \, ind_\ell$

2.5.4 *Projection of future performance*

Moore's law states that the number of transistors on a chip doubles roughly every two years. How far this will extend into the future is a subject of debate, but assuming Moore's law, and assuming that a doubling of transistors is roughly equivalent to a doubling of the number of feasible function calls, and assuming that (by multiple staging) sufficient memory will be available, we can give a rough estimate of the future performance of the PAMIR programs. We have seen that the number of function calls scales with dimension p as 2^{Kp}, with K given in terms of the level count by Eq. (2.26). Letting ΔY be the number of years ahead that Moore's law holds, and $\Delta(Kp)$ the corresponding increase in the feasible product Kp, we get the estimate

$$\Delta(Kp) \simeq \Delta Y/2 \tag{2.28}$$

for the increase in program performance that might be expected ΔY years in the future.

2.6   False positives and their avoidance

A *false positive* occurs when the thinning condition for a subregion is obeyed, and the integral over that subregion is harvested by the program, even though the *actual error*, as opposed to the error estimated by the thinning condition, is large. Any sampling method for evaluating integrals is subject to false positives for functions that take special values (for example, zero) on the sampling points used for the error estimator. A general discussion of the unreliability of sampling methods is given in Sec. 8.1 of Krommer and Ueberhuber (1998). In this section we give examples of false positives in PAMIR, and discuss strategies for dealing with them.

The simplest example of false positives is found by using the direct hypercube programs such as cube579 to evaluate monomial integrals. When thinning is started at level 1 and the order of the monomial is less than or equal to the order of the integration formula used, the iteration terminates at the initial level, and $I_a - I_b$ is of order the truncation error. However, when a monomial is integrated that is of higher order than the integration formula used, the adaptive program does not always start to iterate. For example, using the fifth order hypercube formula in dimension $p = 4$, the program iterates and gives an accurate answer for the integrand x_1^6, but not for the integrand $x_1^2 x_2^2 x_3^2$. The reason is that the latter function, although of higher order than that of the integration formula, vanishes on the hyperplanes spanning the axes where the fifth order integration formula samples the integrand, and so the I_a and I_b evaluations give the same answer (zero), and the thinning condition is obeyed for arbitrarily small ϵ. So in this case a false positive occurs at the initial level of subdivision.

Since the sampling points in the simplex integration formulas are on oblique, rather than axis-parallel, lines or planes, this problem is not so readily seen with multinomial test functions, but we have nonetheless found examples of false positives in using the simplex programs. For example, using fifth order integration and symmetric subdivision, the $p = 5$ monomial $x_1x_2x_3x_4^2x_5$, when computed with thinning starting at any level below 3, develops a false positive at level 2 and gives an answer that is wrong in the fourth decimal place, even though the output error measures suggest much higher accuracy. Another striking example of a false positive found in using the simplex integration formulas is given in the discussion of the two loop self-energy master function in Sec. 3.4.

There are several general ways to guard against false positives. The simplest is to use the freedom of choosing the parameter *ithinlev* to delay thinning until several subdivisions have taken place. False positives are most dangerous if they occur in the initial few levels, since these have the largest subregions, and if a subregion is prematurely harvested, there is a possibility of significant error. On the other hand, thinning becomes most important after several subdivisions have taken place, when the number of subregions is large. So there can be a useful tradeoff between starting thinning early and starting it late. A good procedure is to test for the stability of the final answer as *ithinlev* is increased; digits that are not stable when *ithinlev* is varied cannot be trusted. (Similarly, one should look for stability of the answer as the error parameter ϵ is decreased.) If computer time permits, one can even do a check by choosing *ithinlev* greater than the limit on the number of levels, which suppresses thinning altogether, and gives the approximate Riemann sum corresponding to the level of subdivision attained.

A second general way to guard against false positives is to compute the integral using alternative options, for example, using integration programs of several different orders, or where allowed as an option for simplex integrals, to use recursive instead of symmetric subdivision. In the fifth order $p = 5$ example noted above, changing to seventh order integration, or changing from symmetric to recursive subdivision while maintaining fifth order integration, both eliminate the false positive at level two.

A third way is to add a function with known integral to the integrand, which has significantly different local behavior, and to subtract its known integral from the total at the end. For example, in the hypercube case, consider the integral

$$0 = \int_{-1}^{1} \phi_q(x), \quad \phi_q(x) = \frac{1}{(q+x)^2} - \frac{1}{q^2-1}, \quad (2.29)$$

which exists for any $q > 1$. Adding a multiple of

$$\prod_{\ell=1}^{p} \phi_q(x_\ell) \quad (2.30)$$

to the test monomial integrands does not change the expected answer, but forces the adaptive program to start to subdivide at level 1 (for small enough ϵ) in all monomial cases. It is of course not necessary for the added function to have an integral that can be evaluated in closed form. In the $p = 5$ simplex case discussed above, we eliminated the false positive at level 2 by numerically integrating the function $(1+x_1)^{-1}$, and then adding a multiple of this function to the integrand and subtracting its integral from the answer. When adding such an auxiliary function, it is probably a good idea to rescale it so that its order of magnitude is similar to that of the integral being evaluated. Clearly there is an infinite variety of such auxiliary functions that can be added to the integrand, each of which shifts the false positive problem to a different part of integrand function space. Even when one is dealing with generic integrands, in which the program starts to subdivide as expected, adding such functions will alter the pattern of subdivision, and can be used (in addition to changing the integration formula parameters) to give further estimates of the errors in the output values $I_{a,b}$ provided by the integration algorithm.

We do not recommend just changing the integration formula parameters as a way of eliminating false positives. The reason is that the samplings in both the simplex and hypercube cases take place on hyperplanes that are determined by the general structure of the integration formulas, but do not vary as the parameters in the integration formulas are changed. So if a false positive is associated with a zero or constant integrand value on one of these hyperplanes, it will not be eliminated by changing the parameter values. Similar remarks apply to changing the thinning function as a way of eliminating false positives.

We have not written into the programs another way of creating a criterion for thinning, the comparison of results from integration programs of different orders (say, of fifth and seventh order). In the simplex example discussed above, doing this would eliminate the false positive, since the seventh order routine uses sampling points that avoid the problematic hyperplanes sampled by the fifth order routine. However, in this case one may as well do two seventh order samplings to set up the thinning condition, and thus benefit from the higher accuracy accruing from use of the seventh order routine for smooth integrands.

2.7 Programming remarks

So far in describing how to use the PAMIR programs, we have focused on the main programs files and their use. The subroutine packages do not have to be accessed by the user in normal operation of the adaptive programs. In this section we describe some of the features of programming of the subroutines, for those users who are interested in looking "inside the box". But we again caution that if the subroutines packages are accessed to alter the programs, we strongly recommend doing several test integrals before and after making changes, to make sure they still operate correctly.

(1) *Integer(2) lattice and memory allocation*

In order to conserve memory, the labeling of simplex and hypercube points uses a lattice built on integer(2) arithmetic. This allows 14 levels of subdivision in the initial stage, since $2^{14}=16384$, which is half the maximum integer representable in integer(2). In order to go beyond 14 levels of subdivision in one stage, say to 30 levels of subdivision, one would have to replace 16384 in the subroutines by $2^{30} = 1{,}073{,}741{,}824$, which is half the maximum integer representable in integer(4), replace all integer(2) data type declarations by integer(4), and enlarge the level number limits in the programs. These level number limits correspond to the requirement that the minimum integer(2) lattice spacing must not be smaller than 1, since in integer arithmetic $1/2$ is replaced by 0. For example, in the first stage of subdivision there can be at most 14 levels that pass on subdivided regions to a subsequent level, or including a final level that does not subdivide, a limit of 15 levels overall. An exception to this rule is in the simplex4 programs, where there is an explicit division by 2 in the programs, and so the corresponding limits are 13 and 14. Note that in integer(2) arithmetic, $16384/2 + 16384/2 = 16384 \neq (16384 + 16384)/2 = (-32768)/2 = -16384$, which is why in the simplex4 integration and simplex subdivision programs we have not regrouped added terms into parentheses.

The recirculating and MPI programs make use of the observation that symmetric (or recursive) subdivision of standard simplexes, symmetric and recursive subdivision of Kuhn simplexes, and hypercube subdivision, all give after ℓ subdivisions a subregion that fits within a hypercube of side $1/2^{\ell}$ (or in the recursive case, $1/2^{\ell-1}$). This observation, which is an unproved conjecture supported by our numerical results in the case of standard simplexes, permits a doubling of the number of levels attainable within integer(2) arithmetic in the "r" and "m" programs, as follows. At the start of the second stage of subdivision, each subregion is translated by a shift vector and is rescaled by a factor which expands it to just fit within the initial lattice containing the base region. This permits another 15 (or for recursive subdivision 14) levels in the second stage (with corresponding limits in the simplex4 programs of 14 (or 13)), and so the "r" and "m" programs can subdivide to subregions that have a dimension $2^{-28} = 3.725 \times 10^{-9}$ of the base region dimension. Whether this can be attained in practice for a given dimension p of course depends on available machine memory. Subdivision limits appropriate to the various cases have been incorporated into the main programs.

Because simplex points are represented in integer(2) arithmetic, in order to apply the simplex subroutines to a starting simplex that does not have only 0s or 1s in the vertex coordinates (for example, an equilateral triangle), one would have to change the integer(2) data type declarations to real(4) for the programs to work correctly. This change increases the memory requirements; a better way is to use the mapping of a general simplex to the standard simplex given in Appendix D.

Finally, we note that again with the aim of conserving memory, we have used allocatable memory to store subregion information, allocating memory where needed at each level of subdivision, and de-allocating memory when no longer used. The scheme used for this is shown in Fig. 2.4(a). Each level of the program is processed by the same module, which receives an input array labeled "i" and sends an output array labeled "o" to the next level. In the process of copying the output array from one level to the input array for the next level, both arrays are temporarily present, which accounts for the two terms in square brackets in the second column of Table 2.7. An alternative way of writing the program, which has not been implemented in the current version, would proceed as in Fig. 2.4(b). In this scheme there are two program modules, one for the odd level numbers and one for the even, with the even level module taking an input labeled "e" and sending an output labeled "o" to the odd level module, and with the odd level module taking an input labeled "o" and sending an output labeled "e" to the even level module. In this case, no copying of a large array from one allocatable array to another with a different name is required, and some memory is saved. Rewriting the programs according to this scheme could be considered in the future. However, we note that in running the two-stage "r" and "m" versions of the current programs, we always found the limiting factor to be the running time needed, not the memory space required, so with current hardware implementations changing from the scheme of Fig. 2.4(a) to that of Fig. 2.4(b) would not result in a large gain in performance.

(2) *Real(16) versions*

The programs in simplex579_16, cube357_16, and cube579_16 are real(16) re-writings of those in simplex579, cube357, and cube579 respectively. The real(16) versions are obtained from the corresponding real(8) programs by making the following global substitutions: (1) Replace "d0" by "q0", (2) replace "implicit real(8)"

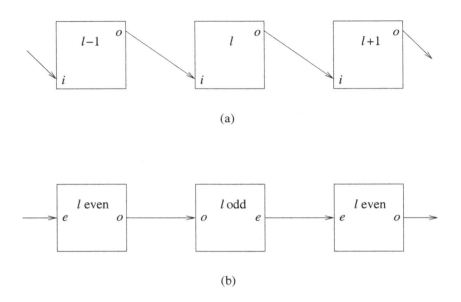

Fig. 2.4 (a) Subregion flow from level to level as programmed. (b) Alternate that would avoid copying one large array to another.

by "implicit real(16)", (3) replace "dabs" by "qabs", (4) replace "d20.13" by "d32.26". These changes were made using a "replace all" utility, since the strings that have to be modified do not occur anywhere else in the programs. Note that explicit data type declarations that override the implicit ones are *not* changed, with the exceptions that in the subroutine fthin, one replaces "real(8)" by "real(16)", and in the MPI main programs, one adds a declaration "real(8) timestart, timefinish". The real(16) MPI programs give only real(8) output, because they use a MPI_DOUBLE specification for variables passed between processes. (This is one of the reasons why the explicit real(8) declarations are not modified in the conversion substitutions leading to real(16) programs.) Nevertheless, the MPI programs compute the sensitive parts of the high order integrations in real(16), converting to real(8) only at the end when process outputs are combined. For full quadruple precision MPI, the specifications MPI_DOUBLE would have to be changed to MPI_LONG_DOUBLE, but we have not tried this.

(3) *MPI version structure*

The MPI programs are written using only simple MPI_Send and MPI_Recv commands. All processes simultaneously carry out the first stage of subdivision, and then each process of rank greater than 0 takes its share of the remaining subregions after the first stage and processes them further. This wastes some processor time, but avoids large data transfers. Only at the end, when all processes of rank greater than 0 have finished, is their output combined in process 0. An alternative way of programming the MPI versions, using the MPI_REDUCE command in place of MPI_Send and MPI_Recv, is given as an illustrative example in the program simplexmain579m1.for in the folder loop_examples.

(4) *Handling standard and Kuhn simplexes*

The simplex programs all perform adaptive integration over a standard simplex or a Kuhn simplex with one vertex at the origin. The same adaptive program treats both the standard and Kuhn simplex cases, with a subroutine argument "i_init" determining which initialization is used.

(5) *Naming of sampling parameters*

The program variable names in array constructors in the "setparam" subroutines, and in the corresponding higher order integration subroutines in the subroutine packages, have been chosen to roughly correspond to the symbol names in the formulas of Chapter 5. For example, for the direct cube routines, where λ is a free parameter, it is called *aalamb*; in the order 7 routine for simplex integration, λ_1^i and λ_2^i are the respective elements of the array constructors *alamb1* (*blamb1*) and *alamb2* (*blamb2*) corresponding to the first (second) choice of parameter values. However, in the fifth order cube and simplex routines, where only one pair of array constructors is needed, they are called *alamb* (*blamb*), even though the corresponding quantity is labeled λ_1^i in Chapter 5.

(6) *Vandermonde solvers*

The programs internally compute integration weights by solving sets of Vandermonde equations. All programs are self-contained, since their subroutine packages include Vandermonde solvers that compute the explicit solution of the Vandermonde systems of the orders needed. Because the ninth order simplex integration routines and associated Vandermonde equations involve large numbers in computing coefficients, use of real(16) is recommended if one wants to get answers with real(8) accuracy. Solving the Vandermonde equations to get the integration weights need be done only once before adaptive integration begins; this happens in the subroutines with names beginning with "ext", the output of which is then fed to the integration programs that are used repeatedly in the adaptive integration process.

(7) *Using PAMIR to construct f by integration over a function g*

As currently configured, PAMIR cannot be used to construct the function f by integration over another function g, before the further integration of f by PAMIR. The most straightforward way to implement this would be to turn the main program being used to integrate g (which need not be the same as the one used to integrate f) into a subroutine, which would be called from within the subroutine defining f. This only requires access to the main program files, without changing the subroutine packages, but in general would necessitate re-computation of the integration weights used to integrate g every time f is called. Avoiding this inefficiency would require reprogramming involving the subroutine packages.

(8) *Miscellaneous comments*

As noted in Sec. 1.2, the programs use Fortran 77 conventions for comment lines and "do" loops, while using a number of Fortran 90 constructs. To enhance readability of the programs, we have used indents to show the different levels of "if" chains, except in one place in the MPI programs, where we have given the "if", "else if", and "end if" lines statement numbers 97, 98, 99. We have not indented the contents of "do" loops, since in the conventions used these always begin and end with a statement number, and never with an unnumbered "enddo". (The one exception to this is in the subroutine BestLex used for the symmetrization step in the Kuhn tiling programs for integration over hypercubes, which has been taken verbatim from H. D. Knoble's (1995) website.) This is of course a matter of taste; our feeling was that indenting both the "if" chains and "do" loops would result in so many levels of indents that readability of the programs would be decreased.

We also note that the direct hypercube programs were written by minimal modification of the simplex programs, changing array arguments where needed (e.g., "$ip + 1$" for simplexes becomes "$2 * ip$" for hypercubes), but not changing array names. So the array names in the direct hypercube subroutines are not the ones that would naturally be chosen if these programs were written without reference to the simplex case.

Chapter 3

Benchmark examples and comparisons

3.1 Introduction

We give in this chapter a number of examples of integrals done by PAMIR, many in comparison with other integration programs.

For comparisons with higher order deterministic methods, we compare PAMIR to CUBPACK (http://nines.cs.kuleuven.be/software/CUBPACK/), using the file simplexpapertest.f90 in the Examples folder of CUBPACK, which "contains the full example of the paper A. Genz and R. Cools An adaptive numerical cubature algorithm for simplices" (Genz and Cools, 2003). We take FUN(0) = 1.d0 and FUN(1) = fcn, with fcn the function to be integrated, so that the answer is the volume normalized integral over the simplex. For hypercube integrals, we use the tiling method, with the same permutation generator BestLex used in the PAMIR programs. A feature of the CUBPACK results shown in the tables below is that the error estimator is very conservative, generally being one or more orders of magnitude greater than the actual error.

For comparisons with Monte Carlo methods, we compare PAMIR with the programs VEGAS (invented by Lepage(1978)) and MISER as given in Press et al. (1992), (http://apps.nrbook.com/fortran/index.html).

In all PAMIR calculations described below, we took the thinning function option *ithinfun* = 1, and for simplexes, used the simplex subdivision choice *isubdivision* = 1. Other parameter choices will be listed as relevant. In all calculations we used the default values of the sampling parameters in the setparam subroutines.

We suggest that the reader skim through all of the examples given below. In some, but not all, we have given the code needed to incorporate the example function into the program. The tables show that the error estimator $|outdiff|$ generally tracks with the accuracy of the result, but can range from somewhat larger than the actual error to up to a factor of 100 smaller, depending on the function being integrated. A useful rule of thumb is to estimate the upper bound on the error as roughly 10 to 100 times the value of $|outdiff|$. In some of the tables we also show the value of $errsum$, the sum of absolute values of subregion error estimators; this is always greater than $|outdiff|$, with equality only when the error estimators all have constant sign. In many cases $errsum$ and $|outdiff|$ bracket the actual error above and below respectively, but $errsum$ can be an overly conservative estimate of the error.

There is no substitute for checking the stability of an answer with respect to changes in the number of levels of subdivision (*llim* or *llim1* + *llim2*), the thinning switch *ithinlev*, the order of integration *iaccuracy*, and the error parameter *eps*. One can also check the stability of an answer with respect to changes in the sampling parameters for higher order integration that are set by the array constructors in the setparam subroutines; if doing this, the folder "readme" should first be consulted.

3.2 Lattice Green's Function near the branch cut: comparison of PAMIR with CUBPACK

References: Katsura et al. (1971); Kaprzyk and Mijnarends (1986); Methfessel, Boon and Mueller (1983); Morita and Horiguchi (1971a,b).

This is an example of a three-dimensional integral over a Kuhn simplex. The lattice Green's function for various cubic lattices is given by

$$G(E) = \frac{1}{V} \int_{\text{Kuhn simplex}} dxdydz \frac{1}{E - \omega(\pi x, \pi y, \pi z)} , \qquad (3.1)$$

with $1/V = 6$ and with

$$\omega(x, y, z) = \cos x + \cos y + \cos z, \quad \text{simple cubic},$$
$$= \cos x \cos y \cos z, \quad \text{body centered cubic},$$
$$= \cos y \cos z + \cos z \cos x + \cos x \cos y, \quad \text{face centered cubic}.$$
$$(3.2)$$

We will focus in our comparison on the simple cubic case, since this can be reduced to a one-dimensional integral in terms of Bessel functions,

$$G(E) = \frac{1}{\pi^3} \int_0^\pi dx \int_0^\pi dy \int_0^\pi dz (-i) \int_0^\infty dt e^{it(E - \cos x - \cos y - \cos z)}$$

$$= -i \int_0^\infty dt e^{iEt} [J_0(t)]^3, \tag{3.3}$$

with

$$J_0(t) = \frac{1}{\pi} \int_0^\pi dx e^{-it \cos x}. \tag{3.4}$$

We will be interested in the complex point $E = E_R + E_I$, with E_R in the range $1 \leq E_R \leq 3$ and with $E_I = 0.001$, for which the imaginary part of the Green's function gives a good approximation to the imaginary part on the branch cut running along the real axis from $E_R = 1$ to 3. For this value of E_I, the one-dimensional representation becomes

$$(-G_I, G_R) = \int_0^\infty dt e^{-0.001\,t} (\cos E_R t, \sin E_R t) [J_0(t)]^3. \tag{3.5}$$

Setting $t = 30{,}000v$, an integral over v from 0 to 1 is suppressed at the upper limit by $e^{-0.001 \times 30{,}000} \simeq 10^{-13}$, which suffices to give 12 digit accuracy. We do this one-dimensional integral with PAMIR, using the program cubetilemain579r.for linked to cubesubs579.for, with parameters $ip = 1$, $llim1 = 14$, $llim2 = 15$, $eps = 10^{-12}$, $isubdivision = 1$, $iaccuracy = 7$, $ithinlev = 12$, and $ithinfun = 1$. The one-dimensional integrand is input through the function fcn1, as follows:

```
      function  fcn1(ip,x)
c function  to  be  integrated  over  side  1  hypercube  goes  here
      implicit  real(8)  (a-h,m-z)
```

```
      implicit integer (i-l)
      dimension x(1:ip)
c insert function to be integrated in place of the following
c test example
c output is integral of fcn1 over base region, divided by
c base region volume
      fac=30000.0d0
      t=fac*x(1)
c er is the value of the real part of E; we sampled
c 1, 1.5, 2, 2.5
      er=1.5d0
c the next two lines give the integrand respectively
c for the real and -imaginary parts of the Green's
c function, with one or the other commented out
c fcn1=dsin(er*t)*dexp(-.001d0 *t)*(DBESJ0(t))**3*fac
      fcn1=dcos(er*t)*dexp(-.001d0 *t)*(DBESJ0(t))**3*fac
      return
      end
```

This, in a few seconds or less on a personal computer, gives high accuracy answers to which the PAMIR and CUBPACK evaluations of the three-dimensional integral can be compared.

We return now to the three-dimensional integral for the simple cubic lattice Green's function, given by

$$(G_R, -G_I) = \frac{1}{V} \int_{\text{Kuhn simplex}} d^3x \frac{(E_R - \sum_{i=1}^{3} \cos \pi x_i, E_I)}{(E_R - \sum_{i=1}^{3} \cos \pi x_i)^2 + E_I^2}. \tag{3.6}$$

We evaluate this by PAMIR using simpexmain579r.for linked to simplexsubs579.for with $ip = 3$, $isimplex = 2$ (which chooses a Kuhn simplex), $llim1 = 4$, $llim2 = 8$, $eps = 10^{-4}$, $iaccuracy = 5$, $isubdivision = 1$, and $ithinlev = 4$. The three-dimensional integrand is now input through the function fcn, as follows:

```
      function fcn(ip,x)
c function to be integrated over standard simplex goes here
      implicit real(8)(a-h,m-z)
      implicit integer(i-l)
      dimension x(1:ip)
c insert function to be integrated in place of the following
c test example
```

```
c output is integral of fcn over base region , divided by base
c region volume
      pi =3.141592653589793d0
c er is the real part of E; we sampled 1, 1.5,
c 2, 2.5; ei is the imaginary part of E
      er =2.5d0
      ei =.001d0
      sum =0.d0
      do 7 i =1,3
      sum=sum+dcos ( pi*x ( i ))
    7 continue
c the next two lines give the integrand respectively
c for the real and −imaginary parts of the Green's
c function , with one or the other commented out
c fcn =(er−sum)/(( er−sum)**2+ei**2)
      fcn=ei /(( er−sum)**2+ ei **2)
      return
      end
```

This gives the PAMIR results for the real and imaginary parts of the Green's function near the branch cut, as done by evaluating the three-dimensional integral. These, together with the corresponding CUBPACK results, computed with relative error parameter $EpsRel = 10^{-4}$, are given in Tables 3.1 and 3.2.

In both tables the column labeled |actual error| is the difference between the program output (*outav* for PAMIR, Value for CUB-PACK) and the "exact" result obtained from the one-dimensional integral representation. We see that the error measure |*outdiff*| for PAMIR typically underestimates the actual error by an order of magnitude, while the sum of subregion error magnitudes errsum is an order of magnitude larger than the actual error. For CUBPACK, on the other hand, the estimated error (the relative error parameter *EpsRel* times the value of the integral) overestimates the actual error by more than an order of magnitude in the cases where the error goal is attained. We see from these tables that CUPBACK is somewhat, but not decisively, better than PAMIR for evaluating G_R near the branch cut, but CUBPACK has difficulty giving a good answer for the highly peaked integral $-G_I$ near the cut. In all cases the CUB-PACK calculation of $-G_I$ is abandoned (with running times greater

Table 3.1 Simple cubic lattice Green's function at $E_I = 0.001$, $E_R = 1, 1.5, 2, 2.5$. "exact" is answer from one-dimensional integral representation; *outav* and remaining columns are from evaluation of the three-dimensional integral by PAMIR, using simpexmain579r.for linked to simplexsubs579.for, with $ip = 3$, $isimplex = 2$ (which chooses a Kuhn simplex), $llim1 = 4$, $llim2 = 8$, $eps = 10^{-4}$, $iaccuracy = 5$, $isubdivision = 1$, and $ithinlev = 4$. The running time t_final is the time in seconds on a MacBook Pro.

function, E_R	"exact"	outav	$|outdiff|$	errsum	|actual error|	fcncalls	t_final
$G_R, 1$	0.62781147909	0.627832	3.2e-6	5.4e-4	2.1e-5	1.8e9	525
$G_R, 1.5$	0.59803236634	0.598020	1.6e-7	3.2e-4	1.2e-5	1.1e9	322
$G_R, 2$	0.56185140167	0.561847	2.4e-7	1.2e-4	4.4e-6	6.8e8	191
$G_R, 2.5$	0.53137587243	0.531372	3.1e-7	1.8e-5	3.9e-6	2.9e8	82
$-G_I, 1$	0.89417203917	0.894183	3.5e-6	5.5e-4	1.1e-5	1.9e9	542
$-G_I, 1.5$	0.46362430870	0.463612	1.5e-6	3.2e-4	1.2e-5	1.2e9	338
$-G_I, 2$	0.30405998275	0.304054	1.1e-6	1.2e-4	6.0e-6	7.3e8	208
$-G_I, 2.5$	0.18234113857	0.182337	3.9e-7	1.7e-5	4.1e-6	3.5e8	100

Table 3.2 Simple cubic lattice Green's function at $E_I = 0.001$, $E_R = 1, 1.5, 2, 2.5$. "exact" is answer from one-dimensional integral representation; Value and remaining columns are from evaluation of the three-dimensional integral by CUBPACK, with $EpsRel = 10^{-4}$, and t_final is the time in seconds on a MacBook Pro. (†) indicates that the function calls limit of 2e9 was reached, and so the error goal was not attained.

function, E_R	"exact"	Value	est. error	|actual error|	fcncalls	t_final
$G_R, 1$	0.62781147909	0.627812	6.3e-5	1.e-6	2.1e8	118
$G_R, 1.5$	0.59803236634	0.598035	4.5e-4 (†)	3.e-6	2.e9	1185
$G_R, 2$	0.56185140167	0.561853	5.6e-5	2.e-6	1.5e8	85
$G_R, 2.5$	0.53137587243	0.531379	5.3e-5	3.e-6	1.8e7	10
$-G_I, 1$	0.89417203917	0.86644	0.112 (†)	2.7e-2	2e9	1168
$-G_I, 1.5$	0.46362430870	0.43808	0.0995 (†)	2.6e-2	2e9	1184
$-G_I, 2$	0.30405998275	0.30174	0.011 (†)	2.3e-3	2e9	1186
$-G_I, 2.5$	0.18234413857	0.18181	0.00342 (†)	5.3e-4	2e9	1183

than those for PAMIR) before the error goal is attained.

The maximum running time used by PAMIR for any of the entries in the table, which as noted all used the serial routine simplexmain579r.for linked to simpexsubs579.for, was less than 10 minutes on a MacBook Pro. This means that on a modest cluster of 64 or 128 processes, the MPI version of PAMIR (simplexmain579m.for linked to simplexsubs579m.for) has the capability of doing a high accuracy survey of the real and imaginary parts of the lattice Green's function near the branch cut, directly from the three-dimensional defining integral representation. We will present below, as another of our benchmarks, the results of doing such a survey for the harder, four-dimensional integral defining the two loop master function.

3.3 Double Gaussian in seven dimensions: comparison of PAMIR with VEGAS and MISER

In his paper on VEGAS, Lepage (1978) introduces a test integral consisting of the sum of two spherically symmetric Gaussians equally spaced along the diagonal of a hypercubical integration volume. In terms of the components x_i of $\mathbf{x}$, this test integral is

$$I_p = \frac{1}{2} \left(\frac{1}{a\pi^{1/2}} \right)^p \int_0^1 d^p x [e^{-\sum_{i=1}^p (x_i - 1/3)^2/a^2} + e^{-\sum_{i=1}^p (x_i - 2/3)^2/a^2}],$$

$$(3.7)$$

with $a = 0.1$. In this form I_p can be evaluated by the "cubetile" programs which tile a unit hypercube with Kuhn simplexes. In order to apply the direct hypercube "cube" programs, we make the change of variable $x_i = (1 + y_i)/2$ for each i to rewrite I_p as an integral over a half-side 1 hypercube,

$$I_p = \frac{1}{2} \frac{1}{2^p} \left(\frac{1}{a\pi^{1/2}} \right)^p \int_{-1}^1 d^p y [e^{-\sum_{i=1}^p (y_i + 1/3)^2/(4a^2)}$$

$$+ e^{-\sum_{i=1}^p (y_i - 1/3)^2/(4a^2)}].$$

$$(3.8)$$

In order to get a high accuracy evaluation of I_p for comparison purposes, we note that the two Gaussians contribute equally to I_p (to see this, set $y_i \to -y_i$ in Eq. (3.8)), and each individual Gaussian is the pth power of a one-dimensional integral J, giving

$$I_p = J^p \,,$$

$$J = \frac{1}{2a\pi^{1/2}} \int_{-1}^{1} dy\, e^{-(y+1/3)^2/(4a^2)} \,.$$

$$(3.9)$$

The one-dimensional integral J can be evaluated numerically to 13 digit accuracy by running the "cube" program to a depth of twelve total levels. Running the "r" version of the programs with parameter values $ip = 1$, $llim1 = 5$, $llim2 = 7$, $ithinlev = 12$ (no thinning, which makes the results independent of eps and $ithinfun$), and $iaccuracy = 3, 5, 7$, we get from all three runs the result

$$J = 0.9999987857663 \,.$$

$$(3.10)$$

The statistics for running time in seconds (on a MacBook Pro) and the number of function calls for these runs are given in Table 3.3. Running with $iaccuracy = 1$ gave only 10 place accuracy with 12 levels, but gave 13 place accuracy when 20 levels were used (which took about a second, rather than hundredths of a second). Thus, for this calculation $iaccuracy = 3$ is the most cost-effective program, since computer truncation errors are larger for $iaccuracy = 5, 7$.

Table 3.3 Evaluation of J to 13 place accuracy using 3rd, 5th, 7th order cube routines.

$iaccuracy$	$|outdiff|$	$fncalls$	t_final
3	10^{-16}	0.2×10^5	< 0.02s
5	10^{-15}	0.5×10^5	< 0.02s
7	10^{-13}	0.9×10^5	< 0.02s

Running a program to raise J to powers then gives in Table 3.4 the expected results for I_p, with an uncertainty of 1 in the final decimal place.

We proceed now to directly evaluate the double Gaussian integral in dimension 7, using the PAMIR program cubemain13.for linked to cubesubs13.for, with $iaccuracy = 3$, $llim = 5$, and $ithinlev = 2$, comparing this with evaluations by the Monte Carlo based programs MISER and VEGAS as given in the book of Press et al. (1992).

Table 3.4 Evaluation
of powers of J to give
expected values of I_p to
13 place accuracy.

p	$I_p = J^p$
1	0.9999987857663
2	0.9999975715341
3	0.9999963573033
4	0.9999951430740
5	0.9999939288462
6	0.9999927146199
7	0.9999915003951
8	0.9999902861717
9	0.9999890719498

These calculations were done on an Institute for Advanced Study compute server with a speed 5.2 times faster than that of a MacBook Pro. We did the calculation twice, first a low accuracy run with $eps = 10^{-4}$, and then a high accuracy run with $eps = 10^{-9}$. In both cases we did the calculations with MISER with the same number of function calls as used by PAMIR. For the VEGAS runs, we chose $itmax = 64$, and took $ncall$ to be the number of function calls used by PAMIR divided by $itmax$; the actual number of function calls used by VEGAS ended up around 20% smaller than the number of PAMIR function calls. The results are shown in Tables 3.5 and 3.6.

We conclude from this that the serial version of PAMIR is faster than both MISER and VEGAS for comparable or greater accuracy. Thus, the MPI version of PAMIR will substantially outperform either of these programs.

3.4   Two loop self-energy master function

References: Yuasa et al. (2008); Bauberger and Böhm (1995); Kurihara and Kaneko (2006); Fujimoto et al. (1992); Broadhurst (1990); Kreimer (1991); Passarino and Uccirati (2002).

Table 3.5 Comparison of PAMIR (to left of vertical line) with MISER (to right of vertical line), for evaluation of the seven-dimensional double Gaussian integral. PAMIR used cubemain13.for linked to cubesubs13.for, with *iaccuracy* = 3, *llim* = 5, and *ithinlev* = 2. For both comparisons, the number of function calls for MISER was chosen equal to that for PAMIR. The actual error was calculated by comparison with the result 0.99999150 obtained by raising the one-dimensional integral to the 7th power. Timings are on a compute server that is approximately 5.2 times faster than a MacBook Pro.

| | PAMIR | | | | | MISER | | | | |
eps	outav	|outdiff|	|actual error|	fcncalls	t_final	ave	var	|actual error|	fcncalls	t_final
1.e-4	0.99962	1.6e-6	3.7e-4	2.6e8	37	0.99951	1.9e-7	4.8e-4	2.6e8	60
1.e-9	0.9999914	9.2e-7	1.e-7	1.1e9	161	0.99989	4.5e-8	1.e-4	1.1e9	266

Table 3.6 Comparison of PAMIR (to left of vertical line) with VEGAS (to right of vertical line), for evaluation of the seven-dimensional double Gaussian integral. PAMIR used cubemain13.for, with $iaccuracy = 3$, $llim = 5$, and $ithinlev = 2$. For both comparisons, we took $itmax$ in VEGAS to be 64, and took $ncall$ to be the number of function calls used in PAMIR divided by 64. The actual error was calculated by comparison with the result 0.99999150 obtained by raising the one-dimensional integral to the 7th power. Timings are on a compute server that is approximately 5.2 times faster than a MacBook Pro.

PAMIR						VEGAS					
eps	outav	$|outdiff|$	$|$actual error$|$	fcncalls	t_final	tgral	sd	$|$actual error$|$	ncall	fcncalls	t_final
1.e-4	0.99962	1.6e-6	3.7e-4	2.6e8	37	0.9968	2.2e-3	3.2e-3	4.1e6	2.1e8	95
1.e-9	0.9999914	9.2e-7	1.e-7	1.1e9	161	0.9983	6.8e-4	1.7e-3	1.8e7	9.2e8	414

The master integral for the two loop self-energy takes the form, in the Feynman diagram labeling convention of Yuasa et al.,

$$T_{12345}(p^2, m_1^2, m_2^2, m_3^2, m_4^2, m_5^2) = \int_{\text{standard simplex}} dx_1 \cdots dx_4 \frac{-1}{CD},$$

(3.11)

with

$$C = (x_1 + x_2 + x_3 + x_4)x_5 + (x_1 + x_2)(x_3 + x_4),$$

$$D = -p^2[x_5(x_1 + x_3)(x_2 + x_4) + (x_1 + x_2)x_3x_4 + (x_3 + x_4)x_1x_2]$$
$$+ C\tilde{M}^2,$$

(3.12)

$$\tilde{M}^2 = \sum_{i=1}^{5} x_i m_i^2,$$

$$x_5 = 1 - x_1 - x_2 - x_3 - x_4.$$

Writing $D = D_R + iD_I$, the real and imaginary parts of T_{12345} are given by

$$T_{12345;R,I} = \int_{\text{standard simplex}} dx_1 \cdots dx_4 \frac{-D_R, D_I}{C(D_R^2 + D_I^2)}.$$

(3.13)

Since our simplex program computes the volume normalized integral over the standard simplex, we must include an extra factor $V = 1/24$. Taking $m_1^2 = 1$, $m_2^2 = 2$, $m_3^2 = 3$, $m_4^2 = 4$, $m_5^2 = 5$ in the convention of Bauberger and Böhm, which corresponds to $m_1^2 = 1$, $m_2^2 = 2$, $m_3^2 = 4$, $m_4^2 = 5$, $m_5^2 = 3$ in the Feynman diagram labelling convention of Yuasa et al., the subroutine fcn becomes:

```
      function fcn(ip,x)
c function to be integrated over standard simplex goes here
      implicit real(8)(a-h,m-z)
      implicit integer(i-l)
      dimension x(1:ip)
      common/step/psqr
c insert function to be integrated in place of the following
c test example
c output is integral of fcn over base region, divided by base
c region volume
      m1sq=1.d0
      m2sq=2.d0
```

```
        m3sq=4.d0
        m4sq=5.d0
        m5sq=3.d0
c  psqr  is  the  real  part  of  p-squared;  and  is  input
c  from  main  through  common/step/psqr
c  psqi  is  the  imaginary  part  of  p-squared
        psqi =0.001d0
        x5=1.d0-x(1)-x(2)-x(3)-x(4)
        mmsq=m1sq*x(1)+m2sq*x(2)+m3sq*x(3)+m4sq*x(4)+m5sq*x5
        cc=(x(1)+x(2)+x(3)+x(4))*x5+(x(1)+x(2))*(x(3)+x(4))
        ee=(x5*(x(1)+x(3))*(x(2)+x(4))+(x(1)+x(2))*x(3)*x(4)
       1+(x(3)+x(4))*x(1)*x(2))
        ddr=-psqr*ee+cc*mmsq
        ddi=-psqi*ee
        fac =24.d0
c  the  next  two  lines  give  the  integrand  respectively
c  for  the  real  and  imaginary  parts  of  the  two  loop
c  function ,  with  one  or  the  other  commented  out
c  fcn=(-1.d0/(fac*cc))*(ddr)/(ddr**2+ddi**2)
        fcn =(-1.d0/(fac*cc))*(-ddi)/(ddr**2+ddi**2)
        return
        end
```

With this input function, we proceeded to make a survey of the values of T_R and T_I versus $psqr = p_R^2$, with $psqi = p_I^2$ fixed at 0.001, using the MPI program simplexmain579m.for linked to simplexsubs579m.for. Our initial scan used the parameter values $isimplex = 1$ (which selects a standard simplex), $isubdivision = 1$, $llim1 = llim2 = 5$, $iaccuracy = 5$, $ithinlev = 3$ and $eps = 1.e-4$. Surveying 50 values with p_R^2 ranging in steps of 1 from 1 to 50 took 75 minutes for T_R, and 54 minutes for T_I, on a 64 process cluster. This survey showed a very clear example of a *false positive*, our term for what happens when the two samplings of a subregion used to make a local error estimate are accidentally within eps of each other, even though the estimate for that subregion is poor, resulting in the integration over that subregion being prematurely harvested, instead of the subregion being passed on for further subdivision. Scanning the survey, all values of T_R and T_I showed reasonable continuity, except for a large jump in T_I at $p_R^2 = 41$, as shown by the following successive p_R^2, T_I pairs: $(39, -0.187)$, $(40, -0.177)$, $(41, -0.505)$, $(42, -0.159)$, $(43, -0.149)$. Doing a survey closer to $p_R^2 = 41$ gave the following pairs: $(40.8, -0.168)$, $(40.9, -0.170)$, $(41.0, -0.505)$, $(41.1, -0.166)$, $(41.2, -0.167)$. So there is no evidence of a cusp at 41, and no reason to expect one physically, since the threshold values of p_R^2 corresponding to the chosen masses are $(1 + \sqrt{2})^2 = 5.828$, $(2 + \sqrt{5})^2 = 17.944$, $(1 + \sqrt{3} + \sqrt{5})^2 = 24.682$, and $(2 + \sqrt{2} + \sqrt{3})^2 = 26.484$. The outlier at 41 is evidently a computational artifact.

This was confirmed by changing the values of the integration parameters *ithinlev* and *eps*. As shown in Table 3.7, changing either *ithinlev* or *eps* eliminates the false positive, and the result is stable with respect to further changes in these parameters, and also with respect to increase in the second stage level number *llim2*. This example shows a general feature, that **to be sure one is getting a sensible answer, and to assess the number of significant digits one is getting, it is important to check stability of the answer as** *ithinlev* **and the total level number** *llim1 + llim2* **are increased, and as the error measure** *eps* **is decreased.**

Table 3.7 Values of T_I at $p_R^2 = 41$, $p_I^2 = 0.001$, versus integration parameters, showing false positive and its elimination.

ithinlev	*eps*	*llim2*	T_I
3	1.e-4	5	−0.505
4	1.e-4	5	−0.176
5	1.e-4	5	−0.178
3	1.e-5	5	−0.178
4	1.e-5	6	−0.178

For our final surveys, we used the more conservative parameters *ithinlev* = 6, *llim1* = *llim2* = 6, and *eps* = 1.e − 6, which required longer cluster runs. We present the results in two tables. In Table 3.8, we give values of T_R and T_I for $p_R^2 = 0.1, 0.5, 1.0, 5.0, 10.0, 50.0$ for comparison with the high accuracy results obtained from a one-dimensional integral representation of the master function by Bauberger and Böhm. In the same table, we also give the values of T_R and T_I at the threshold values of p_R^2 listed above. This 10 point survey, on a 128 process cluster, took 8.6 hours for T_R and 5.9 hours for T_I. We then did a 50 point survey of values of T_R and T_I for p_R^2 ranging from 1 to 50 in steps of 1, with results given in Table 3.9 and Fig. 3.1. On a 128 process cluster, this took 70.5 hours for T_R and 47.6 hours for T_I. Based on the comparison with Bauberger and Böhm, we expect the results of this larger survey to be accurate to the one percent level or better.

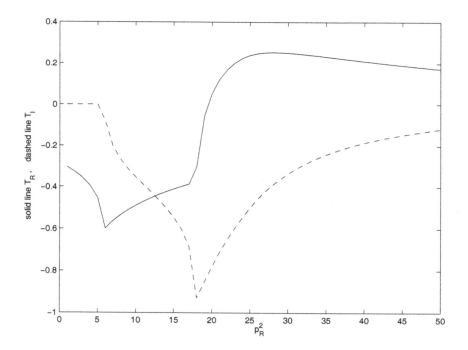

Fig. 3.1   Plot of two loop master function data in Table 3.9; solid line gives T_R, dashed line gives T_I.

Table 3.8 Survey of T_R and T_I versus p_R^2 at $p_I^2 = 0.001$ by direct four-dimensional integration by PAMIR with $llim1 = llim2 = 6$, $iaccuracy = 6$, $ithinlev = 5$, $ithinlev = 6$, $eps = 1.\text{e-}6$, and comparison with Bauberger and Böhm values for $p_I^2 = 0$ (which are used to calculate |actual error|). The final four entries are threshold values. Upper bounds in the T_I column for Bauberger and Böhm are bounds on $|T_I|$.

p_R^2	Bauberger Böhm		T_R from PAMIR				T_I from PAMIR			
	T_R	T_I	outav	|*outdiff*|	|actual error|	*fcncalls*	outav	|*outdiff*|	|actual error|	*fcncalls*
0.1	−0.287238	< 2e-9	−0.28717	4.6e-5	6.8e-5	2.3e9	−0.17781e-4	4.5e-9	—	1.0e9
0.5	−0.294592	< 2e-9	−0.29453	4.8e-5	6.2e-5	2.3e9	−0.18999e-4	5.0e-9	—	1.0e9
1.0	−0.304522	< 2e-9	−0.30445	5.1e-5	7.2e-5	2.3e9	−0.20754e-4	5.8e-9	—	1.0e9
5.0	−0.452521	<3e-9	−0.45234	1.3e-4	1.8e-4	2.6e9	−0.77980e-4	8.5e-8	—	1.0e9
10.0	−0.488154	−0.353218	−0.48836	9.2e-5	2.1e-4	2.2e12	−0.35329	1.8e-4	7.2e-5	1.7e12
50.0	0.173902	−0.118080	0.17385	1.0e-5	5.2e-5	8.5e12	−0.11815	4.2e-4	7.e-5	5.2e12
5.828	—	—	−0.5963	3.2e-3	—	3.0e9	−0.10515e-2	2.7e-3	—	1.1e9
17.944	—	—	−0.3684	1.8e-3	—	1.7e13	−0.92490	1.7e-3	—	1.2e13
24.682	—	—	0.2357	2.4e-4	—	1.8e13	−0.51319	1.8e-4	—	1.3e13
26.484	—	—	0.2514	1.0e-4	—	1.7e13	−0.43886	4.3e-4	—	1.2e13

Table 3.9   Survey of T_R and T_I versus p_R^2 at $p_I^2 = 0.001$ by direct four-dimensional integration by PAMIR with $lliml = llim2 = 6$, $iaccuracy = 5$, $ithinlev = 6$, $eps = 1.e\text{-}6$.

p_R^2	T_R	fcncalls	T_I	fcncalls	p_R^2	T_R	fcncalls	T_I	fcncalls
1	−0.3045	2.3e9	−0.2075e-4	9.9e8	26	0.2486	1.8e13	−0.4572	1.2e13
2	−0.3274	2.3e9	−0.2536e-4	1.0e9	27	0.2528	1.7e13	−0.4201	1.2e13
3	−0.3560	2.4e9	−0.3245e-4	1.0e9	28	0.2542	1.6e13	−0.3880	1.1e13
4	−0.3941	2.5e9	−0.4518e-4	1.0e9	29	0.2534	1.6e13	−0.3593	1.1e13
5	−0.4523	2.6e9	−0.7798e-4	1.0e9	30	0.2514	1.5e13	−0.3342	1.0e13
6	−0.5983	5.2e9	−0.0809	2.9e9	31	0.2488	1.5e13	−0.3111	9.7e12
7	−0.5630	1.6e11	−0.2025	1.2e11	32	0.2450	1.4e13	−0.2912	9.3e12
8	−0.5340	5.9e11	−0.2662	4.6e11	33	0.2413	1.4e13	−0.2725	9.0e12
9	−0.5094	1.3e12	−0.3136	1.0e12	34	0.2377	1.3e13	−0.2564	8.6e12
10	−0.4884	2.2e12	−0.3533	1.7e12	35	0.2329	1.3e13	−0.2416	8.3e12
11	−0.4695	3.4e12	−0.3899	2.7e12	36	0.2288	1.2e13	−0.2278	8.0e12
12	−0.4529	4.9e12	−0.4248	3.7e12	37	0.2245	1.2e13	−0.2151	7.7e12
13	−0.4374	6.5e12	−0.4613	5.0e12	38	0.2201	1.2e13	−0.2041	7.5e12
14	−0.4238	8.3e12	−0.5010	6.3e12	39	0.2157	1.1e13	−0.1936	7.2e12
15	−0.4108	1.0e13	−0.5464	7.7e12	40	0.2113	1.1e13	−0.1841	7.0e12
16	−0.3986	1.2e13	−0.6028	9.3e12	41	0.2077	1.1e13	−0.1756	6.7e12
17	−0.3865	1.5e13	−0.6853	1.1e13	42	0.2032	1.0e13	−0.1669	6.5e12
18	−0.3017	1.7e13	−0.9296	1.3e13	43	0.1990	1.0e13	−0.1591	6.3e12
19	−0.05531	1.9e13	−0.8538	1.4e13	44	0.1952	9.9e12	−0.1522	6.2e12
20	0.05248	2.0e13	−0.7808	1.4e13	45	0.1911	9.6e12	−0.1453	6.0e12
21	0.1217	2.0e13	−0.7132	1.4e13	46	0.1880	9.4e12	−0.1395	5.8e12
22	0.1688	2.0e13	−0.6523	1.4e13	47	0.1842	9.2e12	−0.1336	5.6e12
23	0.2017	1.9e13	−0.5962	1.4e13	48	0.1804	8.9e12	−0.1278	5.5e12
24	0.2248	1.9e13	−0.5453	1.3e13	49	0.1765	8.7e12	−0.1226	5.3e12
25	0.2396	1.8e13	−0.4989	1.3e13	50	0.1738	8.5e12	−0.1181	5.2e12

3.5   Comparison of PAMIR with CUBPACK for polynomial integrals over standard simplex in five dimensions

We consider next polynomial integrals over a standard simplex which vanish quadratically at several of the simplex faces. Letting

$$S = \sum_{i=1}^{5} x_i, \quad \Pi_j = \prod_{i=1}^{j} x_i^2, \tag{3.14}$$

we consider integrands $(1-S)^2\Pi_3$ and $(1-S)^2\Pi_5$. These are of eighth and twelfth degree respectively, and so are not exactly integrated by the seventh order integration routines in PAMIR and CUBPACK. Instead, with seventh order integration both programs have to subdivide the integration regions to achieve high accuracy. The exact answers are given by the multinomial beta function integral,

$$\int_{\text{standard simplex}} dx_1 \cdots dx_p \, (1 - x_1 - x_2 - \cdots - x_p)^{\alpha_0-1} x_1^{\alpha_1-1} \cdots x_p^{\alpha_p-1}$$

$$= \frac{\prod_{a=0}^{p} \Gamma(\alpha_a)}{\Gamma(\sum_{a=0}^{p} \alpha_a)}, \tag{3.15}$$

with Γ the usual gamma function (see the Wikipedia article on Dirichlet distributions). Since the programs compute the volume normalized integral over a standard simplex, for comparison with the program output the answer obtained from this formula must be multiplied by $1/V = p! = 120$ in five dimensions.

The results obtained from PAMIR using simplexmain579.for linked to simplexsubs579.for, with *iaccuracy* $= 7$ and no thinning, and from CUBPACK, are given in Tables 3.10 and 3.11.

Since PAMIR has been run with no thinning, the number of function calls is independent of the structure of the polynomial being integrated, and depends only on the level number limit *llim*. When *llim* is raised from 4 to 5, the accuracy increases, with a corresponding increase in function calls and running time. On the other hand, the number of function calls used by CUBPACK depends strongly on the complexity of the function being integrated. As the integrand becomes more strongly multi-dimensional, the global adaptive method used by CUBPACK requires an increasing computational effort to rank subregion errors in order to decide which subregion should be further subdivided, and the efficiency of the method sharply decreases. We will see this aspect of CUBPACK in numerous examples, and believe that it is characteristic of global adaptive methods quite generally. These methods are good for integrating low dimensional functions embedded in high dimensional spaces, but become less efficient when the function being integrated is highly multidimensional, since as the integrand becomes more rapidly varying,

Table 3.10 Polynomial integrals evaluated by PAMIR using simplexmain579.for, with $iaccuracy = 7$ and no thinning (i.e., $ithinlev \geq 5$). t_final is the running time in seconds on a MacBook Pro.

integrand	exact	llim	outav	$|outdiff|$	$|actual\ error|$	fcncalls	t_final
$(1-S)^2\Pi_3$	$120 \times 2^4/13! = 0.3083336417e\text{-}6$	4	0.308333610e-6	5.0e-15	3.2e-14	8.4e6	1.2
$(1-S)^2\Pi_5$	$120 \times 2^6/17! = 0.2159199171e\text{-}10$	4	0.2159171e-10	3.7e-17	2.9e-16	8.4e6	1.2
$(1-S)^2\Pi_5$	$120 \times 2^6/17! = 0.2159199171e\text{-}10$	5	0.215919904e-10	1.8e-19	1.7e-18	2.7e8	44

Table 3.11 Polynomial integrals evaluated by CUBPACK, with $EpsRel = 10^{-5}$. t_final is the running time in seconds on a MacBook Pro.

integrand	exact	Value	est. error	|actual error|	fcncalls	t_final
$(1-S)^2\Pi_3$	$120 \times 2^4/13 = 0.3083336417$e-6	0.3083333586e-6	3.1e-12	5.5e-14	6.3e6	4.3
$(1-S)^2\Pi_5$	$120 \times 2^6/17 = 0.2159199171$e-10	0.21592122e-10	2.1e-16	1.3e-16	4.9e8	351

at a certain point the effort required by global adaptive methods in deciding which subregion to further subdivide dominates the effort required to execute high order integration in the subregions.

We have not investigated how a global adaptive strategy, combined with full 2^p subdivision, would compare with the local adaptive method and 2^p subdivision used in PAMIR. CUBPACK proceeds by dividing sides rather than whole simplexes at once, so more decisions are required by CUBPACK than would be required by a global adaptive version of PAMIR. However, we suspect that the problem that surfaced in CUBPACK would only be forestalled in the application of a global adaptive method to full 2^p subdivision, but would still appear for highly multi-dimensional functions. We did not pursue this, because we preferred a local adaptive method for reasons of simplicity, particular with regard to writing a parallel MPI version of each program.

3.6 Comparison of PAMIR with CUBPACK for the Feynman–Schwinger integral over standard simplex in five dimensions

As our next test example, we consider integrals constructed from the Feynman–Schwinger formula for combining perturbation theory denominators,

$$\frac{1}{D_0 D_1 \cdots D_p} = p! \int_{\text{standard simplex}} dx_1 \cdots dx_p$$

$$\times \frac{1}{[(1 - x_1 - x_2 - \cdots - x_p)D_0 + x_1 D_1 + \cdots + x_p D_p]^{p+1}}, \quad (3.16)$$

which is proved in Appendix A. We shall apply this formula in five dimensions by taking choices of $D_{0,\ldots,5}$ that give integrands peaked on the boundary hyperplanes of the standard simplex. Taking $D_0 = 1$ and $D_1 = \cdots = D_5 = 0.1$, we get

$$10^5 = 5! \int_{\text{standard simplex}} dx_1 \cdots dx_5 \frac{1}{[1 - 0.9(x_1 + \cdots + x_5)]^6}, \quad (3.17)$$

which gives an integral that is sharply peaked on the diagonal hyperplane $1 = x_1 + \cdots + x_5$ bounding the simplex. Similarly, taking

$D_1 = 1$, and $D_0 = D_2 = \cdots = D_5 = 0.1$ we get

$$10^5 = 5! \int_{\text{standard simplex}} dx_1 \cdots dx_5 \frac{1}{[0.1 + 0.9x_1]^6}, \qquad (3.18)$$

which gives an integral that is sharply peaked on the simplex quadrant boundary $x_1 = 0$, while taking a general $D_j = 1$ and the other Ds to be 0.1 gives an integral that is sharply peaked on the simplex quadrant boundary $x_j = 0$. We shall construct test functions by adding a piece that is peaked on the diagonal hyperplane boundary to successive numbers of pieces that are peaked on quadrant boundaries, as follows:

$$fcn_{1\,\text{term}} = \frac{1}{(1 - 0.9S)^6},$$

$$fcn_{n\,\text{term}} = fcn_{1\,\text{term}} + \sum_{i=1}^{n-1} \frac{1}{(0.1 + 0.9x_i)^6}, \qquad n = 2, \ldots, 5, \quad (3.19)$$

$$S = \sum_{i=1}^{5} x_i.$$

We begin, in Table 3.12, by comparing PAMIR, using simplexmain579.for linked to simplexsubs579.for, with CUBPACK. For PAMIR we used the parameters $isimplex = 1$ (which chooses the standard simplex), $ip = 5$ (which selects five dimensions), $llim = 5$, $iaccuracy = 7$, and no thinning ($ithinlev \geq 5$, which makes the value of eps irrelevant), while for CUBPACK we chose $EpsRel = 10^{-5}$. The exact answer for these integrals is 10^5 times the number of terms.

We see that CUBPACK outperforms PAMIR for functions that are one or two-dimensional functions embedded in the five-dimensional space. For functions that vary in three independent directions the performance difference is small, and for functions varying in four or more independent directions the performance of CUBPACK degenerates significantly, and is worse than that of PAMIR. The running time for PAMIR increases linearly with the number of terms, reflecting just the additional time for function evaluation, whereas that for CUBPACK increases faster than quadratically with the number of terms, reflecting the added overhead of attempting a global comparison of sides to be divided when many subregions are in play.

Table 3.12 Feynman–Schwinger integrals in five dimensions with varying numbers of terms, evaluated by PAMIR using simplexmain579.for linked to slimplexsubs579.for, with $llim = 5$, $iaccuracy = 7$, and no thinning, and by CUBPACK with $EpsRel = 10^{-5}$. t_final is the running time in seconds on a MacBook Pro.

terms	PAMIR					CUBPACK				
	outav	|outdiff|	|actual error|	fcncalls	t_final	Value	est. error	|actual error|	fcncalls	t_final
1	99998.2	0.3	1.8	2.7e8	48	99999.9	1.	0.1	6.8e5	0.56
2	199996.4	0.7	3.6	2.7e8	54	199999.8	2.	0.2	1.5e7	11
3	299994.6	1.0	5.4	2.7e8	62	299999.7	3.	0.3	6.7e7	50
4	399992.8	1.3	7.2	2.7e8	69	399999.7	4.	0.3	2.0e8	162
5	499991.0	1.6	9.0	2.7e8	76	499999.6	5.	0.4	4.8e8	386

In Table 3.13 we give results for PAMIR, using the "hybrid" program simplexmapmain357.for in the cube357 folder, linked to cubesubs357.for, for the same calculations. This program maps the standard simplex into a hypercube, and then does thinning, for the chosen parameter value *iflagopt* = 2, by hyper-rectangular subdivision in only the *ip*1 most rapidly varying directions. Thus it is well suited for integrating lower dimensional functions embedded in high dimensional spaces. The other relevant parameter values are *imc* = 0 (no Monte Carlo option), *idiff* = *2* (use fourth differences for ranking variability along the axes), *llim* = 6, *eps* = 0.1, *iaccuracy* = 7, and *ithinlev* = 2. These calculations were done on a Samsung N120 netbook, which is 3.5 times slower than a MacBook Pro, so for the CUBPACK comparisons on the right-hand side of the table, we have taken the CUBPACK timings on the right-hand side of the previous table and multiplied them by a factor of 3.5, with all other CUBPACK entries the same.

We see that the "hybrid" program, by not subdividing all directions but rather only subdividing the *ip*1 most varying directions, performs better than the direct simplex program, despite the disadvantage of first having to map the standard simplex into a hypercube. For an appropriate choice of *ip*1, the results from the "hybrid" program are competitive or superior to CUBPACK, for all numbers of terms in the integrand.

3.7 Comparisons of PAMIR with CUBPACK for hypercube test integrals in five dimensions

We turn now to a number of comparisons of PAMIR with CUBPACK for test integrals over a five-dimensional hypercube. For PAMIR, we use the direct hypercube routines in the folders cube13 and cube579, and the "hybrid" routines in the folder cube357. For CUBPACK, we initialize to a Kuhn simplex, and then use the standard tiling of a five dimensional hypercube by 5! = 120 Kuhn simplexes, using the permutation generator BestLex in the same way as it is used in the tiling main programs in the folder simplex579. (We also tried CUBPACK using the constant Jacobian map of hypercube into simplex,

Table 3.13 Feynman–Schwinger integrals in five dimensions with varying numbers of terms, evaluated by PAMIR using simplexmapmain357.for linked to cubesubs357.for, with parameters *iflagopt* = 2, *imc* = 0, *idiff* = 2, *llim* = 6, *eps* = 0.1, *iaccuracy* = 7, and *ithinlev* = 2. *t_final* is the running time in seconds on a Samsung N120. The CUBPACK columns are copied from the previous table, with running times scaled up by a factor of 3.5.

terms	PAMIR							CUBPACK				
	llim	ipl	outav	|outdiff|	|actual error|	fcncalls	t_final	Value	est. error	|actual error|	fcncalls	t_final
1	6	1	99999.1	0.03	0.9	2.1e6	2.9	99999.9	1.	0.1	6.8e5	2.0
2	6	2	199998.	5.6	2.	1.2e7	17	199999.8	2.	0.2	1.5e7	39
2	7	2	199999.84	0.04	0.16	2.3e7	34	199999.8	2.	0.2	1.5e7	39
3	7	3	299999.93	0.05	0.07	2.9e8	447	299999.7	3.	0.3	6.6e7	175
3	7	2	299999.3	1.7	0.7	3.8e7	59	299999.7	3.	0.3	6.6e7	175
4	7	4	399999.95	0.05	0.05	3.8e9	6662	399999.7	4.	0.3	2.0e8	567
4	7	3	399999.86	0.28	0.14	3.5e8	567	399999.7	4.	0.3	2.0e8	567
4	7	2	399993.	26	7.	5.0e7	80	399999.7	4.	0.3	2.0e8	567
5	7	3	499999.8	0.49	0.2	4.0e8	694	499999.6	5.	0.4	4.8e8	1351
5	7	2	499986.8	49	13.	5.5e7	94	499999.6	5.	0.4	4.8e8	1351

but for comparable run times this was not as accurate as the tiling method.)

3.7.1 *Double Gaussian*

We consider first the double Gaussian example, already studied in seven dimensions in Sec. 3.3 above. From the table of powers J^p, we see that the expected answer in five dimensions is 0.9999939288462. We begin in Table 3.14 with a study of CUBPACK results versus the relative error parameter *EpsRel*, which since the integral is very close to 1.0 also gives the estimated error bound.

Table 3.14 Double Gaussian by CUBPACK in five dimensions; t_final is the running time in seconds on a MacBook Pro. The function calls counting includes the factor 120 for tiling the hypercube by Kuhn simplexes.

Value	est. error $< EpsRel$	|actual error|	*fcncalls*	t_final
0.99978	1e-2	2.e-4	2.1e7	9.5
0.999983	1e-3	1.1e-5	8.9e7	39
0.9999936	1e-4	3.e-7	4.8e8	208
0.99999390	1e-5	3.e-8	3.1e9	1348
0.999993931	1e-6	2.e-9	2.2e10	9640

For comparison, we integrate the double Gaussian using various PAMIR routines in five dimensions, as shown in Table 3.15. For $ip1 = ip$ and the same values of *iaccuracy* and the other parameters as in cubemain579.for, cubemain357.for performs the same computation as cubemain579.for, as can be verified by comparing the first and second lines of the table.

Table 3.15 Double Gaussian by PAMIR in five dimensions; t_final is the running time in seconds on a MacBook Pro.

program; parameters	outav	$|outdiff|$	$|$actual error$|$	fcncalls	t_final
cubemain579.for; $llim = 5$, $iaccuracy = 7$					
no thinning	0.999993931	1.e-8	3.e-9	4.8e8	239
cubemain357.for; $ip1 = 5$, $llim = 5$, $iaccuracy = 7$					
no thinning	0.999993931	1.e-8	3.e-9	4.8e8	233
$ithinlev = 2$, $eps = 10^{-11}$	0.999993931	1.e-8	3.e-9	2.4e8	119
cubemain357.for; $iflagopt = 2$ $ithinlev = 2$, $eps = 10^{-11}$					
$ip1 = 4$	0.99982	7.4e-4	1.7e-4	6.4e7	32
$ip1 = 3$	0.99942	1.3e-3	5.7e-4	1.8e7	8.8
cubemain357.for; $iflagopt = 3$ $ithinlev = 2$, $eps = 10^{-11}$					
$eps1 = 10^{-5}$	0.999993930	1.e-8	1.e-9	2.0e8	99
$eps1 = 10^{-4}$	0.999993930	1.e-8	1.e-9	2.0e8	97
$eps1 = 10^{-3}$	0.9999905	2.8e-5	3.4e-6	1.0e8	49
$eps1 = 10^{-2}$	0.999985	4.7e-5	8.9e-6	5.4e7	26
$eps1 = 10^{-1}$	1.00038	1.1e-4	3.9e-4	1.3e7	6.6

For *iflagopt* = 2, cubemain357.for subdivides only the *ip*1 most rapidly varying directions, and for *iflagopt* = 3, cubemain357.for subdivides only those directions that have difference measures greater than *eps*1 times the largest difference measure. These tests were run before introducing the option of varying *idiff*, and so all have *idiff* = 4, corresponding to an eighth order difference measure, the maximum allowable by the program for *iaccuracy* = 7. We see from this comparison with CUBPACK that at low accuracies the performances of PAMIR and CUBPACK are comparable, but at high accuracies PAMIR outperforms CUBPACK by a factor of order 100.

3.7.2   *Tsuda functions*

The Tsuda function is based on the integral

$$\int_0^1 dy \frac{a(a+1)}{(a+y)^2} = 1 . \tag{3.20}$$

In five dimensions this generalizes to the completely symmetric test function

$$fcn = \prod_{i=1}^5 \left(\frac{a(a+1)}{(a+x_i)^2} \right) , \tag{3.21}$$

which integrates to unity over the five-dimensional unit hypercube for any value of a. We will take $a = 0.1$ in the tests that follow. This integrand is input to cubemain357r.for through the function fcn:

```
      function fcn(ip,x)
c function to be integrated over half-side 1 hypercube goes here
      implicit real(8)(a-h,m-z)
      implicit integer(i-l)
      dimension x(1:ip),y(1:ip)
c insert function to be integrated in place of the following
c test example
c output is integral of fcn over base region, divided by base
c region volume
      do 5 i=1,ip
      y(i)=.5d0*(1.d0+x(i))
    5 continue
      a=0.1d0
      fcn=(a*(a+1)/(a+y(1))**2)*(a*(a+1)/(a+y(2))**2)*(a*(a+1)/(a+y(3))**2)
```

```
1*(a*(a+1)/(a+y(4))**2)*(a*(a+1)/(a+y(5))**2)
return
end
```

For the CUBPACK comparison, this function is input to a sub-routine fcn1, which is then symmetrized over all arguments to give the function integrated over a Kuhn simplex by CUBPACK.

Results for the Tsuda function are given in Table 3.16, which shows that CUBPACK outperforms PAMIR by around a factor of 2 at high accuracies. The symmetric Tsuda function is well-suited to integration by CUBPACK, since all Kuhn simplex tilings of the hypercube have the same function, so when these are superposed there is no added multi-dimensional variability.

To see what happens when the integrand is not completely symmetric, we examine an asymmetric Tsuda function constructed as follows. Let

$$f(j_1, j_2, j_3, j_4, j_5) = \prod_{i=1}^{5} \left(\frac{a(a+1)}{(a + x_{j_i}(i))^2} \right) , \qquad (3.22)$$

with $j_1, \ldots, j_5$ taking the values either 0 or 1, and with $x_0(i) = x_i$ and $x_1(i) = 1 - x_i$. For any values of $j_1, \ldots, j_5$, the integral of $f(j_1, j_2, j_3, j_4, j_5)$ over the five-dimensional hypercube is still unity, since we have simply interchanged the roles of the integration limits 0 and 1 for some of the axes. We now define six functions of increasing complexity, as follows:

$$\begin{aligned}
fcn_1 &= f(0,0,0,0,0) , \\
fcn_2 &= fcn_1 + f(1,1,1,1,1) , \\
fcn_3 &= fcn_2 + f(0,1,0,0,0) , \\
fcn_4 &= fcn_3 + f(0,1,1,0,0) , \\
fcn_5 &= fcn_4 + f(0,1,1,1,0) , \\
fcn_6 &= fcn_5 + f(0,1,1,1,1) .
\end{aligned} \qquad (3.23)$$

For example, with $a = 0.1$, the program fcn corresponding to fcn_4 is as follows:

Table 3.16 Tsuda function integrated by PAMIR and CUBPACK. We have grouped on the same line PAMIR and CUBPACK runs of similar accuracies. For PAMIR, we use cubemain357r.for linked to cubesubs357.for, with $iflagopt = 3$, $imc = 0$, $ithinlev = 2$, $llim1 = 2$, $llim2 = 4$, $idiff = 2$, and $eps = 10^{-7}$. For the first PAMIR line, $iaccuracy = 5$, and for the second and third lines, $iaccuracy = 7$. For CUBPACK, since the integral is close to 1.0, the "est. error" is the value of the error parameter *EpsRel*. The function calls counting for CUBPACK includes the factor 120 for tiling the hypercube by Kuhn simplexes. *t_final* is the running time in seconds on a MacBook Pro.

| | PAMIR cubemain357r.for | | | | | | CUBPACK | | | |
eps1	*outav*	|*outdiff*|	|actual error|	*fcncalls*	*t_final*		Value	est. error	|actual error|	*fcncalls*	*t_final*
0.1	0.999965	6.3e-5	3.5e-5	7.1e7	18		0.99985	1.e-3	1.5e-4	1.1e7	1.9
0.1	0.9999985	1.7e-6	1.5e-6	2.0e8	53		0.9999915	1.e-4	8.5e-6	9.3e7	16
0.01	0.99999923	1.3e-6	7.7e-7	9.8e8	255		0.99999959	1.e-5	4.1e-7	7.4e8	136

```
      function fcn(ip,x)
c function to be integrated over half-side 1 hypercube goes here
      implicit real(8)(a-h,m-z)
      implicit integer(i-l)
      dimension x(1:ip),y(1:ip)
c insert function to be integrated in place of the following
c test example
c output is integral of fcn over base region, divided by base
c region volume
      do 5 i=1,ip
      y(i)=.5d0*(1.d0+x(i))
    5 continue
      a=0.1d0
      fcn=(a*(a+1)/(a+y(1))**2)*(a*(a+1)/(a+y(2))**2)*(a*(a+1)
     1/(a+y(3))**2)*(a*(a+1)/(a+y(4))**2)*(a*(a+1)/(a+y(5))**2)
      fcn=fcn+(a*(a+1)/(a+1.d0-y(1))**2)*(a*(a+1)/(a+1.d0-y(2))**2)
     1*(a*(a+1)/(a+1.d0-y(3))**2)
     2*(a*(a+1)/(a+1.d0-y(4))**2)*(a*(a+1)/(a+1.d0-y(5))**2)
      fcn=fcn+(a*(a+1)/(a+y(1))**2)*(a*(a+1)/(a+1.d0-y(2))**2)
     1*(a*(a+1)/(a+y(3))**2)
     2*(a*(a+1)/(a+y(4))**2)*(a*(a+1)/(a+y(5))**2)
      fcn=fcn+(a*(a+1)/(a+y(1))**2)*(a*(a+1)/(a+1.d0-y(2))**2)
     1*(a*(a+1)/(a+1.d0-y(3))**2)
     2*(a*(a+1)/(a+y(4))**2)*(a*(a+1)/(a+y(5))**2)
      return
      end
```

Results of running PAMIR and CUBPACK for these six functions are given in Table 3.17. For PAMIR we used cubemain357r.for linked to cubesubs357.for, with parameters $iflagopt = 3$, $imc = 0$, $llim1 = 2$, $llim2 = 3$, $eps = 10^{-7}$, $iaccuracy = 5$, $ithinlev = 2$, and $eps1 = 0.1$, which gives relative errors in the answer of about 2 parts in 10^3. For the CUBPACK runs, we used $EpsRel = 10^{-3}$, which gives an error bound of 1 part in 10^3 and actual errors of about 2 parts in 10^4. The results in the table show that for PAMIR the running time increases linearly with the complexity of the Tsuda function being integrated, whereas for CUBPACK the increase in running time is faster than quadratic. As in the case of the Feynman–Schwinger integrals discussed above, we attribute this to the extra computational overhead associated with the global adaptive method used by

CUBPACK when a function with complicated multi-dimensional variation is integrated.

3.7.3 *Eighth order polynomial*

As our next example, we consider the eighth order polynomial

$$fcn = x_1^2 x_2^2 x_3^2 x_4^2 , \qquad (3.24)$$

with integral over the unit hypercube $1/3^4 = 0.012345679012345\ldots$. For PAMIR we used cubemain579.for linked to cubesubs579.for, giving results in comparison with CUPACK as shown in Tables 3.18 and 3.19.

We see that PAMIR substantially outperforms CUBPACK at high accuracies.

3.7.4 *Single Gaussian with generic center in five dimensions*

We consider a single Gaussian with a generic center (that is, with a center that is not at the center of the unit hypercube),

$$fcn = \prod_{i=1}^{5} e^{-(x_i - \beta_i)^2 / a^2} , \qquad (3.25)$$

with $a = 0.1$, and with

$$\beta_1 = 0.17, \quad \beta_2 = 0.43, \quad \beta_3 = 0.67, \quad \beta_4 = 0.29, \quad \beta_5 = 0.84 . \quad (3.26)$$

A high accuracy answer is obtained by doing the one-dimensional integrals individually by PAMIR, giving

$$\int_{\text{hypercube}} dx_1 \cdots dx_5 \, fcn = 0.1758088518859$$
$$\times \, 0.1772453849848 \times 0.1772451141081$$
$$\times \, 0.1772417428859 \times 0.1751493151395$$
$$= 0.1714606218 \times 10^{-3} . \qquad (3.27)$$

For PAMIR we use cubemain579r.for linked to cubesubs579.for, with results in comparison with CUBPACK as shown in Tables 3.20 and 3.21.

Table 3.17 Multi-term Tsuda functions integrated by PAMIR and CUBPACK. PAMIR used cube-main357r.for linked to cubesubs357.for, with parameters $iflagopt = 3$, $imc = 0$, $llim1 = 2$, $llim2 = 3$, $eps = 10^{-7}$, $iaccuracy = 5$, $ithinlev = 2$, and $eps1 = 0.1$. CUBPACK used the parameter $EpsRel = 10^{-3}$. t_final is the time in seconds on a Dell Latitude D600, which is about a factor of 2 slower than a MacBook Pro.

		PAMIR				CUBPACK			
n	function	$outav$	$fcncalls$	t_final	$1 + 2n$	$Value$	$fcncalls$	t_final	$2.1 + 2.4n^{2.4}$
1	fcn_1	0.999	7.2e6	3.0	3	0.9999	1.1e7	4.5	4.5
2	fcn_2	1.997	1.1e7	5.6	5	1.9997	2.3e7	11.6	14.8
3	fcn_3	2.996	1.2e7	7.3	7	2.9995	5.6e7	35.1	35.6
4	fcn_4	3.995	1.2e7	9.1	9	3.9993	9.8e7	73.8	69.0
5	fcn_5	4.994	1.3e7	10.7	11	4.9993	1.4e8	115.4	116.3
6	fcn_6	5.992	1.3e7	12.5	13	5.9991	1.9e8	176.5	179.0

Table 3.18 Eighth order polynomial evaluated by PAMIR, using cubemain579.for linked to cubesubs579.for, with *iaccuracy* = 7 and no thinning (*ithinlev* $\geq$ 5). *t_final* is the time in seconds on a MacBook Pro.

parameters	*outav*	$|outdiff|$	$|$actual error$|$	*fcncalls*	*t_final*
llim = 4, *eps* = 10^{-9}	0.0123456790095	1.6e-14	2.8e-12	1.6e7	2.4
llim = 5, *eps* = 10^{-11}	0.01234567901234	3.1e-14	<1.e-14	5.0e8	77

Table 3.19   Eighth order polynomial evaluated by CUBPACK, with the three lines corresponding to $EpsRel = 10^{-4}$, 10^{-6}, and 10^{-8} respectively. t_final is the time in seconds on a MacBook Pro. The function calls counting includes the factor 120 for tiling the hypercube by Kuhn simplexes.

Value	est. error	|actual error|	*fncalls*	t_final
0.0123456694	8.8e-7	9.6e-9	2.3e5	0.069
0.01234567887	1.2e-8	1.4e-10	2.7e7	3.0
0.0123456790121	1.2e-10	2.e-13	1.8e9	191

Again, we see that PAMIR outperforms CUBPACK when high accuracies are required.

3.7.5   *Lorentzians in five dimensions: corner peak, edge peak, generic internal peak*

We consider the product Lorentzian function

$$fcn = \prod_{i=1}^{5} \frac{1}{\alpha_i^2 + (x_i - \beta_i)^2}, \tag{3.28}$$

which for $0 \le \beta_i \le 1$ and $0 < \alpha_i$ has the integral over the five-dimensional hypercube

$$\int_{\text{hypercube}} dx_1 \cdots dx_5\, fcn = \prod_{i=1}^{5} \frac{1}{\alpha_i} \left[\tan^{-1}\left(\frac{1-\beta_i}{\alpha_i}\right) + \tan^{-1}\left(\frac{\beta_i}{\alpha_i}\right) \right]. \tag{3.29}$$

We take $\alpha_i = 0.1$, $i = 1, \ldots, 5$, and as examples of corner peak, edge peak, and generic internal peak Lorentzians, we take the parameters β_i as follows:

$$\text{corner peak}: \quad \beta_1 = 0,\ \beta_2 = 1,\ \beta_3 = 0,\ \beta_4 = 1,$$
$$\beta_5 = 1,$$
$$\text{edge peak}: \quad \beta_1 = 0,\ \beta_2 = 1,\ \beta_3 = 0,\ \beta_4 = 1,$$
$$\beta_5 = 0.23,$$
$$\text{generic internal peak}: \quad \beta_1 = 0.15,\ \beta_2 = 0.76,\ \beta_3 = 0.41\,.\ \beta_4 = 0.87,$$
$$\beta_5 = 0.23\,. \tag{3.30}$$

Table 3.20 Single Gaussian with generic center evaluated by PAMIR, using cubemain579.for linked to cubesubs579.for, with $iaccuracy = 5$ and $ithinlev = 2$. t_final is the time in seconds on a MacBook Pro.

parameters	outav	$|outdiff|$	$|$actual error$|$	fcncalls	t_final
$llim1 = 3$, $llim2 = 3$, $eps = 10^{-10}$	0.1714604305e-3	2.6e-11	1.9e-10	1.8e8	48
$llim1 = 3$, $llim2 = 4$, $eps = 10^{-11}$	0.1714606096e-3	6.2e-13	1.2e-11	4.4e9	1167

Table 3.21 Single Gaussian with generic center evaluated by CUBPACK, with the three lines corresponding to $EpsRel = 10^{-2}$, 10^{-3}, and 10^{-4} respectively. t_final is the time in seconds on a MacBook Pro. The function calls counting includes the factor 120 for tiling the hypercube by Kuhn simplexes.

Value	est. error	|actual error|	$fncalls$	t_final
0.17141e-3	1.7e-6	5.1e-8	1.4e8	31
0.17146055e-3	1.7e-7	7.2e-11	8.9e8	194
0.17146074e-3	1.7e-8	1.2e-10	6.1e9	1322

Table 3.22 Lorentzians in five dimensions integrated by PAMIR, using cubemain579r.for linked to cubesubs579.for, with parameter values $llim1 = llim2 = 3$, $iaccuracy = 5$, $ithinlev = 2$, and eps as shown in the table. t_final is the time in seconds on a MacBook Pro.

type	eps	$outav$	|$outdiff$|	|actual error|	$fncalls$	t_final
corner peak	1.0	689,051.633	2.2e-2	9.6e-2	1.2e9	217
edge peak	1.0	1,218,882.369	2.1e-2	1.0e-1	1.5e9	274
internal peak	100.0	10,755,834.7	0.10	9.2	1.3e9	232

The "exact" answers for these three cases, calculated from the arctan formuala, are

$$\begin{aligned}
\text{corner peak}: \quad & 689{,}051.729\,, \\
\text{edge peak}: \quad & 1{,}218{,}882.470\,, \\
\text{generic internal peak}: \quad & 10{,}755{,}843.94\,.
\end{aligned} \tag{3.31}$$

For PAMIR we use cubemain579r.for linked to cubesubs579.for, with results in comparison with CUBPACK as shown in Tables 3.22 and 3.23. These show that PAMIR is slightly better than CUBPACK for similar run times.

Table 3.23 Lorentzians in five dimensions integrated by CUBPACK, with various values of *EpsRel*. *t_final* is the time in seconds on a MacBook Pro. The function calls counting includes the factor 120 for tiling the hypercube by Kuhn simplexes.

type	*EpsRel*	Value	est. error	|actual error|	*fcncalls*	*t_final*
corner peak	10^{-4}	689,055.0	68	3.3	5.5e8	68
corner peak	10^{-5}	689,051.737	6.9	8.e-3	4.9e9	604
edge peak	10^{-4}	1,218,884.6	122	2.1	1.6e9	202
edge peak	10^{-5}	1,218,882.52	12	5.e-2	1.4e10	1666
internal peak	10^{-3}	10,755,716	10754	128	1.1e9	141
internal peak	10^{-4}	10,755,845	1075	1.1	8.5e9	1073

3.7.6 *Oscillatory cosine integrals in five dimensions*

We consider next the integrand

$$fcn = \cos\left(\beta + \sum_{j=1}^{5} \alpha_j x_j\right) = \text{Re}[e^{i(\beta + \sum_{j=1}^{5} \alpha_j x_j)}], \qquad (3.32)$$

which when integrated over the unit hypercube gives

$$\int_{\text{hypercube}} dx_1 \cdots dx_5 \, fcn = \text{Re}\left[e^{i\beta} \prod_{j=1}^{5} \frac{e^{i\alpha_j} - 1}{i\alpha_j}\right]. \qquad (3.33)$$

(Here we denote components of **x** by x_j rather than by x_i to avoid confusion with the imaginary unit i.) We consider two specific examples, a "slow" cosine

$$fcn = \cos(\pi(1.0 + 0.5x_1 + 1.5x_2 + 2.5x_3 + 1.0x_4 - 1.0x_5)), \qquad (3.34)$$

$\int_{\text{hypercube}} dx_1 \cdots dx_5 \, fcn = \frac{64}{15\pi^5} = 0.0139424582$,

and a "fast" cosine

$$fcn = \cos(\pi(1.0 + 6.5x_1 + 7.5x_2 + 8.5x_3 + 7.0x_4 - 7.0x_5)),$$

$$\int_{\text{hypercube}} dx_1 \cdots dx_5 \, fcn = \frac{64}{162,435\,\pi^5} = 0.128751115 \times 10^{-5}. \qquad (3.35)$$

Results for PAMIR, using cubemain579.for linked to cube-subs579.for, and for CUBPACK, are shown in Tables 3.24 and 3.25. For the "slow" cosine, both PAMIR and CUBPACK give good answers, with PAMIR about twice as fast. For the "fast" cosine, PAMIR gives good answers, but CUBPACK, even with *EpsRel* = 1, failed to give an answer in an hour of computing time. This again illustrates the problems encountered by the global adaptive strategy of CUBPACK in dealing with rapidly varying multi-dimensional integrands.

3.7.7 C_0 *function in five dimensions*

So far all the examples considered have been differentiable to all orders, for which high order integration formulas (such as the fifth and seventh order formulas of PAMIR and CUBPACK) offer considerable advantages of accuracy. We turn now to the consideration of a C_0 (that is, a continuous but not everywhere differentiable) example,

$$fcn = e^{-\sum_{i=1}^{5} |x_i - \beta_i|/a_i} . \tag{3.36}$$

For $0 \leq \beta \leq 1$, the one-dimensional integral

$$\int_0^1 dx\, e^{-|x-\beta|/a} = a[2 - e^{-\beta/a} - e^{-(1-\beta)/a}] \tag{3.37}$$

allows us to evaluate the integral of *fcn* over the hypercube,

$$\int_{\text{hypercube}} dx_1 \cdots dx_5\, fcn = \prod_{i=1}^{5} a_i[2 - e^{-\beta_i/a_i} - e^{-(1-\beta_i)/a_i}]. \tag{3.38}$$

For our example *fcn*, we take

$$a = 0.1, \quad \beta_1 = 0.15, \quad \beta_2 = 0.76, \quad \beta_3 = 0.41, \quad \beta_4 = 0.87, \quad \beta_5 = 0.23, \tag{3.39}$$

Table 3.24 Cosine integrals evaluated by PAMIR, using cubemain579.for linked to cubesubs579.for, with *ithinlev* = 2 and other parameters as shown in the table. *t_final* is the time in seconds on a MacBook Pro.

function	llim	iaccuracy	eps	outav	|outdiff|	|actual error|	fcncalls	t_final
"slow"	5	5	10^{-6}	0.0013942436	3.3e-9	5.4e-9	1.0e8	21
"fast"	5	5	10^{-10}	0.13050×10^{-5}	1.6e-9	1.8e-8	1.0e8	21
"fast"	6	5	10^{-12}	0.1287721×10^{-5}	1.8e-11	2.1e-10	3.3e9	673
"fast"	6	7	10^{-12}	$0.12875069 \times 10^{-5}$	6.9e-13	4.3e-12	1.6e10	3621

Table 3.25   Cosine integrals evaluated by CUBPACK. t_final is the time in seconds on a MacBook Pro. The function calls counting includes the factor 120 for tiling the hypercube by Kuhn simplexes.

function	$EpsRel$	Value	est. error	|actual error|	$fcncalls$	t_final
"slow"	10^{-3}	0.01394235	1.4e-5	1.1e-7	4.1e7	6.0
"slow"	10^{-4}	0.013942455	1.4e-6	3.e-9	3.4e8	49
"fast"	1	no answer	—	—	—	$t > 3600$

which gives for the integral

$$\int_{hypercube} dx_1 \cdots dx_5 \, fcn = 10^{-5}(2 - e^{-1.5} - e^{-8.5})(2 - e^{-7.6} - e^{-2.4})$$
$$\times (2 - e^{-4.1} - e^{-5.9})(2 - e^{-8.7} - e^{-1.3})$$
$$\times (2 - e^{-2.3} - e^{-7.7})$$
$$= 0.220362273922 \times 10^{-3}. \tag{3.40}$$

The integrand is input through the function fcn as follows:

```
      function fcn(ip,x)
c function to be integrated over half-side 1 hypercube goes here
      implicit real(8)(a-h,m-z)
      implicit integer(i-l)
      dimension x(1:ip),y(1:ip)
c insert function to be integrated in place of the following
c test example
c output is integral of fcn over base region, divided by base
c region volume
      do 5 i=1,ip
      y(i)=.5d0*(1.d0+x(i))
5     continue
      a=0.1d0
      sum=dabs(y(1)-.15d0)+dabs(y(2)-.76d0)+dabs(y(3)-.41d0)
1+dabs(y(4)-.87d0)+dabs(y(5)-.23d0)
      fcn=dexp(-sum/a)
      return
      end
```

We compare three different methods of evaluating the integral of *fcn* over the five-dimensional hypercube. In the first, we use the

PAMIR program cubemain123r.for linked to cubesubs123.for, with *ithinlev* = 2 and the other parameters as shown in Table 3.26. In the second, we use CUBPACK, which gives somewhat, but not dramatically, better results than the PAMIR cube programs, as shown in Table 3.27. Neither of these methods gives highly accurate results, which motivated us to develop the "hybrid" program in which a higher order method is employed in regions where the integrand is sufficiently continuous, but the program automatically switches to Monte Carlo in subregions where the integrand is discontinuous.

The "hybrid" method is illustrated in Table 3.28, where we use the PAMIR "hybrid" programs cubemain357.for or cubemain357r.for linked to cubesubs357.for, with *iaccuracy* = 5 and *idiff* = 2 (fourth order differences to assess variation along axes), and with the Monte Carlo option *imc* = 2 chosen along with *iflagopt* = 2. For each subregion, these options choose Monte Carlo when *delta* times the Monte Carlo error estimate (the difference of two samplings of *ihits* each) is less than the error estimate given by the absolute value of the difference between two fifth order integrations with different sampling parameters. When this inequality is not satisfied, the fifth order routine estimates are chosen. The subregion is harvested if the error estimate for the integration method chosen is less than *eps*, and otherwise the subregion is further subdivided by halving the axes with the *ip*1 largest variances, as calculated by fourth order differences. We see that the "hybrid" routine is capable of higher accuracies, for comparable run times, than either the PAMIR hypercube routine or CUBPACK. (We have checked that when *imc* = 2 is replaced by *imc* = 0, which suppresses the Monte Carlo option, the accuracy depreciates sharply; in other words, fifth order integration by itself cannot give the results obtained by the "hybrid" program.) We also note from this table that although *errsum* and |*outdiff*| often bracket the actual error, in some cases, like this one, *errsum* can give a very conservative upper bound on the error.

Table 3.26 Evaluation of C_0 integral by PAMIR using cubemain123r.for linked to cubesubs123.for, with *ithinlev* = 2. *t_final* is the time in seconds on a MacBook Pro.

iaccuracy	llim1, llim2	eps	outav	|outdiff|	errsum	|actual error|	fcncalls	t_final
1	3, 3	10^{-6}	0.219300e-3	2.6e-7	1.9e-6	1.1e-6	3.9e7	12
1	3, 4	10^{-8}	0.220671e-3	2.2e-7	8.4e-7	3.1e-7	3.4e9	1067
3	2, 4	10^{-6}	0.219884e-3	2.0e-8	3.3e-6	4.8e-7	1.3e8	83

Table 3.27 Evaluation of C_0 integral by CUBPACK. t_final is the time in seconds on a MacBook Pro. The function calls counting includes the factor 120 for tiling the hypercube by Kuhn simplexes.

EpsRel	Value	est. error	|actual error|	fncalls	t_final
1	0.22139e-3	2.2e-4	1.0e-6	4.0e7	8.8
0.1	0.220195e-3	2.2e-5	1.6e-7	6.5e9	1429

3.8 Attenuation function of radiation from a disk source

This is an example of a high accuracy evaluation of a two-dimensional integral. Setting the disk radius $R = 1$, the attenuation function of radiation from a disk source is given by

$$I = a \int_0^1 du \int_0^1 dv \frac{v}{[v^2 + d^2 + a^2 - 2dv \cos(2\pi u)]^{3/2}} ; \quad (3.41)$$

to restore general R, one replaces $a \to a/R$ and $d \to d/R$. Ezure (2008) gives a table of 5 digit evaluations of I. With PAMIR, it is easy to get much higher accuracy. In Table 3.29, we give a sampling of a and d values, with 10 digit evaluations of the attenuation function I, with an uncertainty of 1 in the final digit. Underneath I, two function call numbers are given, an initial number which gave the 10 digit answer, and in parentheses the larger number corresponding to more restrictive parameters, used to verify the accuracy of the initial answer.

3.9 High accuracy calculations of double Gaussian integral in seven and nine dimensions on a 64 process cluster

We return here to the double Gaussian integral discussed earlier, in which two compact ($a = 0.1$) Gaussians are located at the 1/3 and 2/3 points along a hypercube diagonal. We give in Table 3.30 results for evaluating this by PAMIR in seven and nine dimensions,

Table 3.28 Evaluation of C_0 function by PAMIR using the "hybrid" programs cubemain357.for or cubemain357r.for linked to cubesubs357.for. All rows have $iaccuracy = 5$, $idiff = 2$, $imc = 2$, $iflagopt = 2$, and $ithinlev = 2$, $eps = 10^{-10}$, and $delta = 0.1$. Where the "levels" entry is $llim$, cubemain357.for with $ihit = 13$ was used; where the "levels" entry is $llim1$, $llim2$, cubemain357r.for with $ihit = 53$ was used. The value of $ip1$ is given in the table. The first three lines of the table were run on an Institute for Advanced Study compute server, and the final line on a Dell Latitude 600, with the running times in both cases re-scaled to those for a MacBook Pro to give the quoted t_final.

ip1	levels	outav	|outdiff|	errsum	|actual error|	fcncalls	t_final
1	$llim = 10$	0.22148e-3	1.9e-6	2.5e-5	1.1e-6	1.0e7	7.8
2	$llim = 10$	0.2204453e-3	1.5e-8	5.3e-6	8.3e-8	3.1e8	228
3	$llim1 = 2$, $llim2 = 6$	0.22037920e-3	5.3e-9	2.9e-6	1.7e-8	6.8e8	486
5	$llim1 = 2$, $llim2 = 5$	0.22036281e-3	7.4e-10	1.1e-6	5.4e-10	1.3e10	7576

Table 3.29 PAMIR evaluation of attenuation function I for radiation from a disk source, with initial and verification function calls listed below each value.

$a\downarrow$ $d\rightarrow$	0	1	50	100
0.01	0.9900005000	0.4893611665	0.4001800510e-7	0.5000005624e-11
	2.6e6 (1.7e8)	9.1e5 (2.0e8)	1.8e4 (4.5e6)	1.8e4 (4.5e6)
5	0.1941932431e-1	0.1837829996e-1	0.1971234421e-4	0.2499909065e-8
	1.8e4 (4.5e6)	1.8e4 (4.5e6)	1.8e4 (4.5e6)	1.8e4 (4.5e6)
1000	0.4999996250e-6	0.4999998750e-6	0.4981304720e-6	0.1767767119e-6
	1.8e4 (4.5e6)	1.8e4 (4.5e6)	1.8e4 (4.5e6)	1.8e4 (4.5e6)

Table 3.30   Double Gaussian results using cubemain13m.for linked to cubesubs13m.for, and cubemain579m.for linked to cubesubs579m.for, for dimension $ip = 7$ and $ip = 9$, on a 64 process cluster; $levels = llim1 + llim2$. t_final is the time in seconds.

dimension ip	$iaccuracy$	$levels$	$outav$	$|outdiff|$	$|$actual error$|$	$fcncalls$	t_final
7	3	5	0.999992	9.e-7	5.e-7	7.8e9	11
7	5	5	0.9999913	5.e-7	2.e-7	4.4e10	120
7	7	5	0.99999150	3.e-8	<1.e-8	2.7e11	750
7	1	6	0.999993	4.e-7	1.2e-6	5.2e11	2900
7	3	6	0.99999151	8.e-8	1.e-8	1.0e12	4300
7	5	6	0.999991498	9.e-9	2.e-9	5.6e12	17,000
7	7	6	0.9999915003	1.e-10	<1.e-10	3.5e13	98,000
9	3	5	0.999989	1.e-6	7.e-8	2.5e12	7500
9	5	5	0.9999888	8.e-7	3.e-7	1.7e13	49,000
9	7	5	0.99998907	7.e-8	2.e-9	1.3e14	380,000

using integration routines of accuracies 1, 3, 5, and 7 on a 64 process cluster. (We also ran examples with 9th order accuracy, but these were less accurate than the corresponding 7th order results because of truncation errors, and required much more running time.) Since no thinning is employed in the calculations of this table, its results are representative of what can be accomplished for any integrand with a scale of variation similar to that of the Gaussian peaks. The column |actual error| is obtained by comparison of the PAMIR values with the values 0.9999915003951 for seven dimensions, and 0.9999890719498 in nine dimensions, obtained above from a one dimensional high accuracy calculation raised to the seventh or ninth power.

Chapter 4

Computational integration theory and PAMIR

4.1 Overview: The philosophy of PAMIR

Our aim in this chapter is to discuss the philosophy behind the construction of the PAMIR programs, in relation to the existing literature on computational integration theory and the application of this theory in programs currently in use. We begin with two disclaimers.

First, we do not intend to give a survey of the literature pertaining to computational integration, even as restricted to multidimensional integration. For a relatively recent book surveying the field of computational integration, which includes chapters on multivariate integration and on parallel algorithms, see Krommer and Ueberhuber (1998), which updates and expands on Krommer and Ueberhuber (1994); certain more recent papers will be referred to in the discussion that follows. Our aim here is to focus on those aspects of PAMIR which *differ* from methods currently in use.

Second, we make no claim that PAMIR is a panacea that solves the notorious "curse of dimensionality" that afflicts high dimensional integration. As recognized by the 2^p subdivision scheme that is at the heart of PAMIR, high dimensional integration is exponentially hard as the dimension p becomes large. The PAMIR programs are designed to cope with this hardness for moderate values of p, by efficient use of the memory and speed capabilities of current computers. As noted in the Introduction, and as demonstrated by examples given in Chapter 3, we have obtained good results for strongly locally peaked or rapidly varying integrands in dimensions up to $p = 7$ on a personal computer, and up to $p = 9$ on a modest sized cluster. For a statistical physics problem requiring the computation of a partition function in dimension $p = 1000$, the PAMIR programs are of no use, and traditional importance sampling methods must be employed.

The underlying philosophy behind the construction of the PAMIR programs recognizes that there is no single right or wrong way to construct a numerical integration program. Constructing such programs is basically an exercise in engineering, involving tradeoffs to achieve certain goals, some at the expense of others, just as engineering tradeoffs are involved in the design of automobile engines, computers, and cellular phones. In this chapter, we discuss the philosophy and tradeoffs involved in the following aspects of PAMIR:

- Choice of integration method — higher order integration versus statistical sampling, and their relation to integrand smoothness.
- Choice of integration rule — "efficient" rules versus rules giving multiple sampling options.
- Choice of error estimator — use of lower order rules in error estimators, versus error estimators using two samplings of the same order.
- Choice of subdivision method and decision method — dividing one dimension at a time versus dividing all p simultaneously, local versus global decision schemes, and their relation to the effective dimensionality of the integrand.

4.2 Choice of integration method

Two classes of integration methods figure prominently in the computational integration literature:

- The first are designed for functions which are smooth, in the sense that they are continuous and differentiable to some high order. For these functions, it can be advantageous to use a higher order integration method, which exactly integrates polynomials up to some specified order. For example, in doing a one-dimensional integral it can be advantageous to use a Simpson method, which is third order accurate, as in the McKeeman (1962) adaptive algorithm, as opposed to the first order accurate trapezoidal rule.

- The second are designed for functions which are not necessarily smooth, such as functions with discontinuities in the function or in low order derivatives. For such functions it can be disadvantageous to use a high order integration method. One is generally better off with some type of statistical sampling method, such as variants of the Monte Carlo method, the number theoretic method for producing equidistributed sequences of points, etc.

The main focus of the PAMIR programs is on functions that are smooth, for which higher order integration formulas are developed for integrations over both simplexes and hypercubes. However, in the cube357 folder we also give a hybrid variant, which compares higher order integration with Monte Carlo integration for each subregion, and picks the one which gives the better result according to criteria summarized in Tables 2.2 and 2.3. This program is suited for functions which are smooth in most of their domain, but have boundary surfaces of discontinuity joining regions with smooth behavior.

4.3   Choice of integration rule

We assume now that we are dealing with smooth functions, and address the issue of choosing an appropriate higher order integration rule. Since Gaussian integration in one dimension is the paradigm for the construction of most standard higher order integration rules, we briefly review it. Consider the one-dimensional integral over the interval $(-1, 1)$,

$$I \equiv \frac{1}{2} \int_{-1}^{1} dx\, f(x)\,. \tag{4.1}$$

Let us look for a set of values x_j and weights w_j such that the integral is well approximated by the integration rule

$$I \simeq R_n \equiv \sum_j w_j f(x_j)\,, \tag{4.2}$$

in the sense that when f is a polynomial of degree $n = 2m - 1$ or less, the rule of Eq. (4.2) exactly gives the integral of Eq. (4.1).

As we shall see, there is more than one way to do this. The most efficient way to construct a rule, in the sense of involving a minimum number of sampling points x_j and hence of function calls, is the method of Gaussian integration. Let P_m be a polynomial of degree m such that

$$\int_{-1}^{1} dx\, P_m(x) x^k = 0 \qquad (4.3)$$

for all $k < m$. The theory of orthogonal polynomials tells us that the m zeros of P_m lie in the integration interval $(-1, 1)$, and that in fact P_m is proportional to the order m Legendre polynomial. Let $f(x)$ now be polynomial of degree $2m - 1$ or less. Using polynomial long division, we can write

$$f(x) = P_m(x) q(x) + r(x) , \qquad (4.4)$$

with the quotient $q(x)$ and the remainder $r(x)$ both of degree $m - 1$ or less. Then by Eq. (4.3) we have

$$\int_{-1}^{1} dx\, P_m(x) q(x) = 0 ,$$
$$\frac{1}{2} \int_{-1}^{1} dx\, f(x) = \frac{1}{2} \int_{-1}^{1} r(x) . \qquad (4.5)$$

Substituting Eq. (4.4) into our putative integration rule, we have

$$R_n = \sum_j w_j [P_m(x_j) q(x_j) + r(x_j)] , \qquad (4.6)$$

which if we choose the points x_j to be the m zeros of $P_m(x)$ reduces to

$$R_n = \sum_{j=1}^{m} w_j r(x_j) . \qquad (4.7)$$

Now let us fix the m weights w_j by the requirement that the rule of Eq. (4.7) should be exact for all polynomials of degree less than m, which implies m conditions and so can be satisfied. Since $r(x)$ is of degree less than m, for this choice of weights we have

$$\frac{1}{2} \int_{-1}^{1} dx\, r(x) = \sum_j w_j r(x_j) , \qquad (4.8)$$

which in turn implies by Eq. (4.5) that

$$\frac{1}{2}\int_{-1}^{1} dx\, f(x) = \sum_{j} w_j f(x_j)\,. \tag{4.9}$$

In other words, by choosing m sampling points x_j and weights w_j according to the Gauss procedure, we have gotten an integration rule that is exact for all polynomials of degree $n = 2m - 1$ or less. The zeros of the Legendre polynomials and the associated Gaussian integration weights have been extensively tabulated numerically, and there are also efficient algorithms for computing them.

Let us now study an alternative way of getting a rule that is exact for polynomials of degree $2m - 1$ or lower. We assume that $f(x)$ is power series expandable on the interval $(-1, 1)$, so that we can write

$$f(x) = f_0 + f_1 x + f_2 x^2 + f_3 x^3 + f_4 x^4 + f_5 x^5 + f_6 x^6 + f_7 x^7 + f_8 x^8 + \cdots, \tag{4.10}$$

and again consider the one-dimensional integral

$$I = \frac{1}{2}\int_{-1}^{1} f(x)dx$$
$$= f_0 + \frac{f_2}{3} + \frac{f_4}{5} + \frac{f_6}{7} + \frac{f_8}{9} + \cdots\,. \tag{4.11}$$

Defining $\Sigma_1(\lambda)$ by

$$\Sigma_1(\lambda) = f(\lambda) + f(-\lambda)\,, \quad 0 < \lambda \leq 1\,, \tag{4.12}$$

we look for an integration formula of the form

$$I = \kappa_0 f(0) + \frac{1}{2}\sum_{j=1}^{m-1} \kappa^j \Sigma_1(\lambda^j)$$
$$= \kappa_0 f_0 + \sum_{j=1}^{m-1} \kappa^j [f_0 + f_2(\lambda^j)^2 + f_4(\lambda^j)^4 + f_6(\lambda^j)^6 + f_8(\lambda^j)^8 + \cdots]\,. \tag{4.13}$$

In both κ^j and λ^j, j is a superscript and *not* an exponent, a convention that will be followed in analogous discussions in Chapter 5.

Matching the coefficients of f_0, f_2, f_4, ... between Eqs. (4.11) and (4.13), we get the system of equations

$$1 = \kappa_0 + \sum_{j=1}^{m-1} \kappa^j \,,$$

$$\frac{1}{3} = \sum_{j=1}^{m-1} \kappa^j (\lambda^j)^2 \,,$$

$$\frac{1}{5} = \sum_{j=1}^{m-1} \kappa^j (\lambda^j)^4 \,,$$ \hfill (4.14)

$$\frac{1}{7} = \sum_{j=1}^{m-1} \kappa^j (\lambda^j)^6 \,,$$

$$\frac{1}{9} = \sum_{j=1}^{m-1} \kappa^j (\lambda^j)^8 \,,$$

and similarly if one wishes to go to higher order than ninth.

There are now two ways to proceed to solve the matching equations, to give a discrete approximation to the integral to a given order of accuracy. The first is to regard all of the λ^j as free parameters, and to determine the weights κ^j to satisfy the system of equations of Eq. (4.14) to the needed order, for any values of these parameters. Thus, to get a first order accurate formula, we take $I \simeq f(0)$ with all the κ^j equal to zero and $\kappa_0 = 1$, which is the center-of-bin rule. To get a third order accurate formula we must take κ^1 as nonzero and solve the system

$$1 = \kappa_0 + \kappa^1 \,,$$
$$\frac{1}{3} = \kappa^1 (\lambda^1)^2 \,.$$ \hfill (4.15)

To get a fifth order accurate formula we must take both κ^1 and κ^2 as nonzero and solve the system

$$1 = \kappa_0 + \kappa^1 + \kappa^2 \,,$$
$$\frac{1}{3} = \kappa^1 (\lambda^1)^2 + \kappa^2 (\lambda^2)^2 \,,$$ \hfill (4.16)
$$\frac{1}{5} = \kappa^1 (\lambda^1)^4 + \kappa^2 (\lambda^2)^4 \,,$$

to get a seventh order accurate formula we must take $\kappa^{1,2,3}$ as nonzero and solve the system

$$1 = \kappa_0 + \kappa^1 + \kappa^2 + \kappa^3 \,,$$

$$\frac{1}{3} = \kappa^1 (\lambda^1)^2 + \kappa^2 (\lambda^2)^2 + \kappa^3 (\lambda^3)^2 \,,$$

$$\frac{1}{5} = \kappa^1 (\lambda^1)^4 + \kappa^2 (\lambda^2)^4 + \kappa^3 (\lambda^3)^4 \,, \tag{4.17}$$

$$\frac{1}{7} = \kappa^1 (\lambda^1)^6 + \kappa^2 (\lambda^2)^6 + \kappa^3 (\lambda^3)^6 \,,$$

and so forth. Evidently, to get an order $n = 2m - 1$ formula, we must take $m - 1$ distinct positive sampling points $\lambda^{1,\dots,m-1}$, together with the $m - 1$ sampling points $-\lambda^{1,\dots,m-1}$ that are opposite in sign, and the point 0. To determine the weights $\kappa^1, \dots, \kappa^{m-1}$ we must solve an order $m - 1$ Vandermonde system, consisting of all equations except the initial one involving κ_0. Vandermonde equations typically arise in moment matching problems, and are discussed in more detail in Sec. 5.4. Substituting the solution for $\kappa^1, \dots, \kappa^{m-1}$ into the initial equation then determines κ_0. The resulting order $2m - 1$ integration formula uses $2m - 1$ function values.

An alternative way to proceed is to adjust the values of the sampling points so that fewer of them are needed to satisfy the matching conditions. This is what is done in the Gaussian integration method which we discussed above, which gives a more efficient scheme, in terms of the number of function calls, starting with third order. Referring to Eq. (4.15), we can evidently achieve a third order match by taking

$$\kappa_0 = 0, \quad \kappa^1 = 1, \quad \lambda^1 = \frac{1}{\sqrt{3}} \,. \tag{4.18}$$

Similarly, referring to Eq. (4.16), we can achieve a fifth order match by taking

$$\kappa_0 = \frac{4}{9}, \quad \kappa^1 = \frac{5}{9}, \quad \kappa^2 = 0, \quad \lambda^1 = \frac{\sqrt{3}}{\sqrt{5}} \,. \tag{4.19}$$

Proceeding in this way, we can obtain the general Gaussian integration formula, which for order $2m - 1$ integration involves m points. Of course, the usual derivation of the Gaussian integration rule given above does not proceed this way, but instead uses polynomial long division to relate the special points λ^j to zeros of the Legendre polynomials.

We see that in one dimension, keeping the sampling points arbitrary, rather than giving them special values, requires approximately a doubling in the number of sampling points and function calls. When Gaussian integration was invented, calculations were done by hand, and a decrease in labor of a factor of two was very significant. But from the point of modern computational complexity theory, a factor of two (or any small numerical factor) is not significant. This is almost true from a practical point of view with modern computers, where a factor of two (or a few) typifies the speedup of the computer available now to the one we bought two years ago. (This argument would not apply for a real-time computer application such as aircraft scheduling, where small differences in speed can make a substantial difference in the likelihood of queueing delays. But if you are using a cluster to evaluate a complicated integral, it is almost certain that you are not looking for an answer in real time.) *So our underlying philosophy, with regard to choice of integration rule, is not to care about reducing the number of sampling points to the absolute minimum, as long as only a loss of a small factor (typically 2 or 3 in all of the PAMIR programs) is involved. We instead place a premium on having general sampling points, so that we can achieve redundancy in our evaluation of integrals using an order $n = 2m - 1$ method.*

Following on these one-dimensional preliminaries, we can now discuss what happens when we consider higher order integration formulas in dimensions greater than one. Since in higher dimensions there is no analog of the one-dimensional polynomial long division rule, there is no universal higher dimensional analog of the Gaussian integration rule, but there are a multitude of special formulas using specially chosen sampling points in higher dimensions, designed to reduce the number of sampling points and function calls. An older review is given in Stroud (1971), and a more recent review is given in Krommer and Ueberhuber (1998). The CUBPACK programs, which we have compared with PAMIR using many examples in Chapter 3, make use of specially chosen sampling points. These are entered, together with the corresponding weights, as multi-digit numerical tables.

On the other hand, the method of keeping the sampling points λ^j as free parameters, and solving a set of Vandermonde equations to get the coefficients κ^j, readily extends to higher dimensions, and this is the method used in the PAMIR programs to generate higher order rules for simplexes and hypercubes. In the PAMIR programs for simplexes, one set of sampling points is fixed and the rest are selected by free parameters; in the hypercube programs, all are selected by free parameters. Despite requiring a greater than minimal number of function calls, the method used in PAMIR has, in our view, a number of advantages: (i) As already stressed, by changing the parameters one gets an inequivalent evaluation of the integral to the same order; (ii) Tedious entry of tables of numerical values of sampling points and weights is not needed — these are either free parameters or are determined by the program at the start of execution by solving linear Vandermonde equations, for which a closed form solution and efficient algorithms are available; (iii) Although the number of function calls is not minimal, it still grows only as a polynomial with the dimension p. Moreover, in the PAMIR rules of order $n = 2m - 1 \geq 5$ in dimension p the *leading power* in the number of function calls is $x^{m-1}/(m-1)!$, with $x = p + 1$ for simplexes and $x = 2p$ for hypercubes (see Table 2.6), which is known to be optimal [for simplexes, see Stroud (1971) and Grundmann and Möller (1978); for hypercubes, see Lyness (1965)]; (iv) For simplexes, the PAMIR programs work directly from the vertex coordinates of a simplex in general dimension p; (v) For both simplexes and hypercubes, the PAMIR programs determining the weights are universal in the sense that they apply to all values of p, and their complexity does not change with p.

This last point distinguishes the moment matching method from parameterized integration formulas appearing in the prior literature.

The simplex method of Silvester (1970) introduces a one parameter family of points spanning simplexes, but gives explicit results only for triangles and tetrahedra, not for general p-dimensional simplexes. The hypercube method of Genz (1986) introduces parameters and gives an explicit formula for the weights, which involves multiple sums and products that increase in complexity with the dimension p.

We give now explicit comparisons of the number of function calls required by PAMIR with the number required by competitive programs. For simplexes, we give in Table 4.1 the number of function calls used by PAMIR to give two independent evaluations, in comparison with the number used in the paper of Genz and Cools (2003) to give an evaluation and an error measure. We see that for the fifth and seventh order rules, PAMIR is roughly a factor of 2 less efficient, and for the ninth order rule, PAMIR is roughly a factor of 3 less efficient. For hypercubes, we give in Table 4.2 the number of function calls used by PAMIR for two independent evaluations by a seventh order method, with the number to give two seventh order evaluations by a combination of the rules G_P and G_G given by Genz (1986). We see that the loss of efficiency of the PAMIR rules is at most a factor of 2 in this case, and diminishes as the dimension p increases. Thus, the behavior of the PAMIR moment matching rules in these multi-dimensional cases is not significantly different from what we found in one dimension in our comparison of the moment matching and Gaussian methods.

We conclude this section with a discussion of the role of positivity of weights in proving convergence. There are *two* senses in which one could ask for convergence of a numerical integration method to the required integral. (i) One could choose to not subdivide the initial integration domain, and ask for convergence as the order n of the integration rule R_n approaches infinity. In this case there are theorems (see Krommer and Ueberhuber (1998), Secs. 5.2.2 and 6.2.2, for a review) stating that when the integration weights are non-negative, one can prove convergence of the limit of the nth order rule to the required integral for continuous integrands. (ii) One could keep the order of the integration rule fixed, and ask for convergence as the

Table 4.1   Comparison of PAMIR function calls for generating two samplings by $n = $ 5th, 7th, 9th order rules for simplex integration in dimension p, with the function calls needed for an integral evaluation and error estimate computed by the corresponding rules of Genz and Cools (2003).

n	$p \rightarrow$	2	3	4	5	6	7	8	9
5	PAMIR	31	43	56	70	85	101	118	136
5	Genz and Cools	16	23	31	40	50	61	73	86
7	PAMIR	71	117	176	249	337	441	562	701
7	Genz and Cools	32	49	86	126	176	237	310	396
9	PAMIR	168	316	531	827	1219	1723	2536	3136
9	Genz and Cools	65	114	201	315	470	675	940	1276

Table 4.2   Comparison of PAMIR function calls for generating two samplings by a 7th order rule for hypercube integration in dimension p, with function calls for two samplings computed by the 7th order rules G_P and/or G_G of Genz (1986).

$p \rightarrow$	2	3	4	5	6	7	8	9
PAMIR	69	153	281	461	701	1009	1393	1861
Genz $2G_P$	34	78	162	302	514	814	1218	1742
Genz $G_P + G_G$	38	96	202	372	622	968	1426	2012
Genz $2G_G$	42	114	242	442	730	1122	1634	2282

degree of subdivision of the integration domain approaches infinity, so as to give a Riemann sum in the limit. In this case, convergence is expected irrespective of whether or not the integration weights are non-negative. Both PAMIR, and the CUBPACK programs to which we have compared it, are programs that proceed by repeated subdivision of the integration domain, rather than by increasing the order of the integration rule (which would be impractical in higher dimensions). Because both methods lead to Riemann sums in the limit of fine subdivision, in both PAMIR and CUBPACK the integration weights can be positive or negative.

4.4   Choice of error estimator

In adaptive integration, error estimators are used in two ways. The first is in deciding whether a given subregion should be subdivided further (in local subdivision methods), or in deciding which subregion should be subdivided next (in global subdivision methods). The second is in forming an estimate of the probable error in the final answer, once the subdivision process has terminated. We give here a very brief discussion of error estimators, referring the reader to Sec. 8.1 of Krommer and Ueberhuber (1998) for a detailed pedagogical exposition that addresses many of the issues.

To discuss strategies for error estimation, it is again useful to return to the one-dimensional case of Gaussian integration for the integral $I = \frac{1}{2} \int_{-1}^{1} dx f(x)$. Let R_n be the Gaussian rule of order n for this integral. Since the Gaussian weights w_j are non-negative, the convergence theorems noted at the end of the preceding section imply that for continuous $f(\mathbf{x})$, the limit $lim_{n \to \infty} R_n = I$. At first sight, this would suggest using a "null rule" formed from the difference of successive odd order Gaussian rules, say $|R_n - R_{n-2}|$, as an estimate of the error in R_n. However, this does not work, because of oscillatory behavior (so-called "phase factor effects") in the approach of R_n to the true value I (Lyness and Kaganove, 1976); it is possible for $|R_n - R_{n-2}|$ to be smaller than an error criterion ϵ, while the difference $|R_n - I|$ remains larger than ϵ. To reduce the effects of oscillations in the Gaussian estimates, Berntsen and Espelid (1991) suggested a method that uses a sequence of null rules formed from the difference of rules of different orders. By forming appropriate norms from two or more independent null rules, the effects of phase oscillations can be minimized. This method has been extended to the case of multi-dimensional integrals when one has a sequence of rules R_n employing specified sampling points. The successive differences of these rules of different orders give "null rules", which can then be used in groups to form error estimators of the Berntsen–Espelid type. Typically, to save function calls, one uses a sequence of rules that are "embedded", in the sense that the lower order rules use some subset of the sampling points of the highest order rule. The CUBPACK programs use an error estimator of the Berntsen–Espelid type.

The method just outlined has both advantages and disadvantages. The advantage is that it works in most cases, and gives conservative error estimates. The disadvantages are: (i) It is fairly complicated to implement; (ii) Because lower order rules are part of the error estimator, the procedure does not take full advantage of the accuracy of the original higher order rule used to compute the integral; (iii) Related to (ii), it can give overly conservative error estimates (some of the error estimates for the CUBPACK examples in Chapter 3 are two orders of magnitude larger than the actual error); (iv) It does not solve the problem of "false positives", since a function that vanishes on the embedded set of sampling points automatically gives a zero error estimate, even when having a nonzero integral and nonzero error in the integral. As already noted in Sec. 2.6 on "false positives", a detailed discussion of error estimation methods, and their vulnerabilities, is given in Sec. 8.1 of Krommer and Ueberhuber (1998).

In PAMIR we employ a different strategy, made possible by our use of parameterized integration rules which permit independent samplings of the integrand function over a subregion. *Instead of forming error estimators from null rules constructed as the difference of rules of different order with fixed sampling points, in PAMIR we construct error estimators from null rules constructed as the difference of two rules of the same order, using two different sets of the parameters which determine the sampling points.* For subregion error estimates, one takes the absolute value of this difference; for error estimates for the complete integral, one can either sum the subregion absolute values, or use the difference of the two samplings summed over all subregions without taking absolute values. This strategy also has advantages and disadvantages. A disadvantage is that as an error estimate for the final answer, the version summing without taking absolute values tracks the actual error to within an order of magnitude or so, but is not guaranteed to give an upper bound on the error, while the version summing absolute values of subregion errors is often too conservative (see the examples in Chapter 3). But the considerable advantages are: (i) The strategy is easy to implement — one just uses the parameterized higher order rule twice, with two different sets of parameters, and takes the difference; (ii) One does not waste the accuracy of the higher order rule by incorporating a rule of lower order into the error estimator; (iii) As a consequence of (ii), if in any subregion the integrand is a polynomial of order no higher than the order of the rule, then the estimator gives zero (up to machine truncation errors) and this subregion can be eliminated as a candidate for further subdivision; (iv) If desired, by taking a *different* two sets of sampling parameters, one can get an independent check on the final answer given by the adaptive integration procedure.

4.5   Choice of subdivision method and decision method

We consider next the related issues of the choice of subdivision method, and the method used to decide whether or when to divide a subregion. Most adaptive integration programs to date subdivide only one dimension at a time. A method that is widely used by physicists to evaluate Feynman parameter integrals is the VEGAS program of Lepage (1978), which uses a hypercube as the base geometry. VEGAS is a Monte Carlo method, in which random samplings of the integration volume are done with a separable probability density that is a product of one-dimensional densities along each axis. This probability density is then iterated to give a more detailed sampling along axes on which the projection of the integrand is rapidly varying. The deterministic method of Genz and Cools (2003) is based on simplexes as the base regions. This algorithm picks the subregion with the largest estimated error, and subdivides it into up to four equal volume subregions by cutting edges along which the integrand is most rapidly varying. The Genz and Cools algorithm, and related adaptive algorithms, are discussed in the survey of CUBPACK by Cools and Haegemans (2003). The CUBA set of algorithms described by Hahn (2005) includes both Monte Carlo methods and deterministic methods; the former include refinements of VEGAS and the latter proceed by bisection of the subregion with largest error. A survey of many types of high dimensional integration algorithms, including adaptive algorithms, is contained in the HIntLib Manual of Schürer (2008). All of these methods avoid a full 2^p subdivision as being computationally prohibitive, which until recently has been the case.

The alternative of a 2^p subdivision has been entertained in earlier work. The paper of Cools and Haegemans (2003) notes the possibility of a 2^p subdivision but does not implement it, and an earlier algorithm of Kahaner and Wells (1979) discusses a 2^p subdivision for simplexes, but no implementation details are available in the literature. *Our strategy in PAMIR has been to implement a full 2^p subdivision, to get an adaptive program that can follow local peaks and valleys of multi-dimensional functions.* Recent increases in computer speed, decreases in memory cost, and the advent of cluster architectures make this strategy feasible. Moreover, use of binary number functions, as in Moore's (1992) subdivision method for simplexes, and its analog for hypercubes, make it possible to write simple subroutines implementing 2^p subdivision for these geometries.

Once one has picked a subdivision method, one then has to decide on how to choose which subregions to further subdivide. As reviewed in Sec. 8.2.2 of Krommer and Ueberhuber (1998), two methods are prominent. In a *local subdivision strategy*, the decision on which subregions to divide further is based on a local error criterion within each subregion. Subregions which do not satisfy the criterion are further subdivided, while those that satisfy the criterion are "harvested" to give a contribution to the final answer, and in the terminology of Krommer and Ueberhuber, transfer from being "active" to "inactive". In a *global subdivision strategy*, all subregions remain active, and their error estimates are rank ordered, with the subregion with the highest error estimate being the next chosen for subdivision. Numerical studies comparing local versus global methods for one-dimensional integrals, such as Malcolm and Simpson (1975), favor the global methods, which have been widely used since. For example, CUBPACK, with which we have given extensive comparisons in Chapter 3, uses a global subdivision strategy.

It has been generally assumed that since global methods are better in one dimension, they will necessarily be superior in the multi-dimensional case. However, our comparisons in Chapter 3 of PAMIR, which uses a local subdivision strategy, with CUBPACK show that this is *not* the case. The examples given there show that what is important in deciding which method is better is the effective dimensionality of the function being integrated. If the function is a one or low dimensional function embedded in a high dimensional space, the subdivision method and global strategy of CUBPACK work better. But for truly multi-dimensional functions, with localized peaks and valleys or rapid variations not aligned with the coordinate axes, CUBPACK becomes inefficient. For such functions the local strategy of PAMIR, combined with its full 2^p subdivision, gives better results in our tests on serial machines.

The future of high performance computing lies not in serial machines, but in clusters exploiting massive parallelism. It is feasible to program a global subdivision algorithm for a cluster, as for example in the exploratory studies of Yuasa et al. (2008) and further work cited there. However, to implement a global adaptive method in a parallel setting requires communication between the various processes at every stage of the calculation, and so involves relatively advanced parallel programming. Perhaps for this reason, we have not found any parallel multi-dimensional integration program available for distribution in "off-the-shelf" form. By contrast, *it is relatively easy to turn a local subdivision algorithm, such as PAMIR, into a parallel program*; the different processes can proceed with subdivision independently of one another, and need only communicate when they combine partial integrals to give a final total at the end of the calculation. So for every PAMIR program, we have written a cluster version, using the Message-Passing Interface (MPI) protocol, and these are available in the on-line PAMIR archive. The cluster versions permit a many-fold increase in the power of the PAMIR programs beyond what is possible on a serial computer.

Chapter 5

Details of construction of the PAMIR algorithms and programs

In this chapter we give the theory behind the PAMIR algorithms, both for simplexes and hypercubes, as well as a few illustrative pieces of Fortran code. We first focus on simplex and hypercube properties, and the method used for their subdivision, including the hyperrectangular subdivision variant of the hypercube case. We then turn to the construction, by moment matching, of parameterized higher order integration formulas for simplexes and hypercubes. Since this entails the solution of Vandermonde equations, we briefly discuss Vandermonde solvers. At the end of the chapter, we describe some special features of the hybrid cube357 programs, and conclude by giving the program listings, with added explanatory comments, for the basic adaptive module that appears in all of the PAMIR programs, and for the code that distributes residual subregions to the second stage of the "r" and "m" programs. We close with a short list of possible programming extensions and open mathematical questions pertaining to the algorithms.

5.1 Simplex properties

Any set of $p + 1$ points in p-dimensional space defines a p-simplex, and we will be concerned with integrations over the interior region defined this way. Thus, in 1 dimension, 2 points define a 1-simplex that is the line segment joining them, in 2 dimensions, 3 points define a 2-simplex that is a triangle, in three dimensions, 4 points define a 3-simplex that is a tetrahedron, and so forth. We will refer to the $p + 1$ points, that each define a p-vector, as the vertices of the simplex, and our strategy will be to express all operations, both for the subdivision of simplexes and for calculating approximations to integrals over simplexes, directly in terms of these vertices.

We remind the reader that we use boldface to denote p-vectors, and enumerate the $p + 1$ vertices of a simplex using the letters a, b, c, d which range from 0 to p, as in $\mathbf{x}_a$, $a = 0, \ldots, p$. The p components of a p-vector are set in italic, not boldface, and are enumerated with the letter i which ranges from 1 to p, as in x_i, $i = 1, \ldots, p$. So a general p-vector $\mathbf{x}$ can be written in terms of components as $\mathbf{x} = (x_1, x_2, \ldots, x_p)$. In this notation, the ith component of the ath simplex vertex is denoted by a double subscript x_{ai}, $a = 0, \ldots, p$; $i = 1, \ldots, p$. In the next section, we will also use the notation k_1, k_2 to label simplex vertices, but we will always reserve the letter i for p-vector components.

A simplex forms a convex set. This means that for any integer $n \geq 1$ and any set of points $\mathbf{x}_1, \ldots, \mathbf{x}_n$ lying within (or on the boundary of) a simplex, and any set of non-negative numbers $\alpha_1, \ldots, \alpha_n$ which sum to unity,

$$\alpha_m \geq 0, \quad m = 1, \ldots, n, \quad \sum_{m=1}^{n} \alpha_m = 1, \tag{5.1}$$

the point

$$\mathbf{x} = \sum_{m=1}^{n} \alpha_m \mathbf{x}_m \tag{5.2}$$

also lies within (or on the boundary of) the simplex $\big($see, e.g., Osborne (2001)$\big)$.

In constructing integration rules for simplexes, we will be particularly interested in linear combinations of the form of Eq. (5.2) in which the points $\mathbf{x}_1, \ldots, \mathbf{x}_n$ are vertices of the simplex. For such sums, one can state a rule which determines precisely where the point $\mathbf{x}$ lies with respect to the boundaries of the simplex. Let $\mathbf{x}_0, \mathbf{x}_1, \ldots, \mathbf{x}_p$ be the vertices of a simplex, and let $\mathbf{x}_c$ denote the centroid of the simplex,

$$\mathbf{x}_c = \frac{1}{p+1} \sum_{a=0}^{p} \mathbf{x}_a \,. \tag{5.3}$$

Let us denote by $\tilde{\mathbf{x}}_a$ the vertices referred to the centroid as origin,

$$\tilde{\mathbf{x}}_a = \mathbf{x}_a - \mathbf{x}_c \,, \tag{5.4}$$

which obey the constraint following from Eq. (5.3),

$$\sum_{a=0}^{p} \tilde{\mathbf{x}}_a = 0 \,. \tag{5.5}$$

Correspondingly, let $\mathbf{x}$ denote a general point, and let $\tilde{\mathbf{x}} = \mathbf{x} - \mathbf{x}_c$ denote the general point referred to the centroid as origin. Since we are assuming that the simplex is non-degenerate, the vectors $\tilde{\mathbf{x}}_a$ span a complete basis for the p-dimensional space, and so we can always expand $\tilde{\mathbf{x}}$ as a linear combination of the $\tilde{\mathbf{x}}_a$,

$$\tilde{\mathbf{x}} = \sum_{a=0}^{p} \alpha_a \tilde{\mathbf{x}}_a \,. \tag{5.6}$$

This expansion is not unique, since by Eq. (5.5) we can add a constant c to all of the coefficients α_a, without changing the sum in Eq. (5.6). In particular, we can use this freedom to put the expansion of Eq. (5.6) in a standard form, which we will assume henceforth, in which the sum of the coefficients α_a is unity,

$$\sum_{a=0}^{p} \alpha_a = 1 \,. \tag{5.7}$$

For coefficients (called barycentric coordinates) obeying this unit sum condition, we can use Eqs. (5.3) and (5.4) to also write

$$\mathbf{x} = \sum_{a=0}^{p} \alpha_a \mathbf{x}_a \,. \tag{5.8}$$

In terms of the expansion of Eqs. (5.6) through (5.8) we can now state a rule for determining where the point $\mathbf{x}$ lies with respect to the simplex: (1) If all of the α_a are strictly positive, the point lies inside the boundaries of the simplex; (2) If a coefficient α_a is zero, the point lies on the boundary hyperplane opposite to the vertex $\mathbf{x}_a$, and if several of the α_a vanish, the point lies on the intersection of the corresponding boundary hyperplanes; (3) If any coefficient α_a is negative, the point lies outside the simplex. This rule is stated without proof for general p in Pontryagin (1952) and for triangles ($p = 2$) in the Wikipedia article on barycentric coordinates. A proof of this rule is given in Appendix E.

In constructing integration rules for simplexes, we will use the following elementary corollary of the result that we have just stated. Consider the sum

$$\tilde{\mathbf{X}} = \sum_{a=1}^{N} \lambda_a \tilde{\mathbf{x}}_a \,, \tag{5.9}$$

with the coefficients λ_a obeying

$$\lambda_a > 0, \quad a = 1, \ldots, N, \quad \sum_{a=1}^{N} \lambda_a < 1 \,, \tag{5.10}$$

with the points $\tilde{\mathbf{x}}_a$ any vertices of a simplex. Some vertices may be omitted, and some used more than once, in the sum of Eq. (5.9). Then the point $\tilde{\mathbf{X}}$ lies inside the simplex. To see this, we note that by adding a positive multiple of zero in the form of Eq. (5.5), the sum of Eq. (5.9) can be reduced to the form of Eqs. (5.6) and (5.7), with all expansion coefficients α_a strictly positive. By the rule stated above, this implies that the point $\tilde{\mathbf{X}}$ lies within the simplex.

5.2   Simplex subdivision

5.2.1   *Simplex subdivision algorithms*

Two very simple algorithms for subdividing simplexes have been given in the computer graphics literature by Moore (1992). Let us denote the vertices of the starting simplex by $\mathbf{x}_0, \ldots, \mathbf{x}_p$, each of which is a p-vector, and from these let us form the p-vectors $\mathbf{V}(k_1, k_2)$ defined by

$$\mathbf{V}(k_1, k_2) = \frac{1}{2}(\mathbf{x}_{k_1} + \mathbf{x}_{k_2}), \quad k_1, k_2 = 0, \ldots, p \,. \tag{5.11}$$

Thus, $\mathbf{V}(0,0) = \mathbf{x}_0$, $\mathbf{V}(0,1) = (1/2)(\mathbf{x}_0 + \mathbf{x}_1)$ and so forth, so that the vectors $\mathbf{V}(k_1, k_2)$ consist of the original simplex vertices, together with the midpoints of the original simplex edges. Let $k = 0, \dots, 2^p - 1$ be an index which labels the 2^p subsimplexes into which the original simplex is to be divided. Moore then gives two algorithms, which he terms *recursive subdivision* and *symmetric subdivision*, for determining the vertices to be assigned to the subsimplex labeled with k. Both make use of the binary representation of k, and of a function determined by this representation, the bitcount function $b(k)$, which is the number of 1 bits appearing in the binary representation of k.

The recursive subdivision algorithm proceeds as follows. As the 0 vertex of the subsimplex labeled by k, take the vector $\mathbf{V}\big(b(k), b(k)\big)$, that is, $k_1 = k_2 = b(k)$. To get the other vertices, scan along the binary representation of k from right (the units digit) to left. For each 0 encountered, add 1 to k_2, and for each 1 encountered, subtract 1 from k_1. The sequence of vectors $\mathbf{V}(k_1, k_2)$ obtained this way gives the desired $p + 1$ vertices of the kth subsimplex.

The symmetric subdivision algorithm proceeds as follows. As the 0 vertex of the subsimplex labeled by k, take the vector $\mathbf{V}\big(0, b(k)\big)$, that is, $k_1 = 0$, $k_2 = b(k)$. To get the other vertices, scan along the binary representation of k from right (the units digit) to left. For each 0 encountered, add 1 to k_2, and for each 1 encountered, add 1 to k_1. The sequence of vectors $\mathbf{V}(k_1, k_2)$ obtained this way gives the desired $p + 1$ vertices of the kth subsimplex.

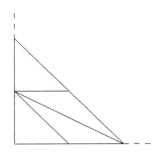

Fig. 5.1   Recursive subdivision of a standard simplex.

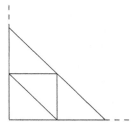

Fig. 5.2   Symmetric subdivision of a standard simplex.

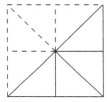

Fig. 5.3   Recursive subdivision of a Kuhn simplex, and the corresponding subdivision of a square tiled with Kuhn simplexes.

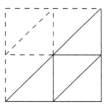

Fig. 5.4   Symmetric subdivision of a Kuhn simplex, and the corresponding subdivision of a square tiled with Kuhn simplexes.

The application of these algorithms in the $p = 2$ case is illustrated in Tables 5.1 and 5.2 and Figs. 5.1–5.4, and in the $p = 3$ case is illustrated in Tables 5.3 and 5.4, where the notations $(k_1, k_2)^{(a)}$ and $\mathbf{x}^{(a)}$ both refer to the ath vertex, $a = 0, \ldots, p$, of the subdivided simplex labeled by the k in each row. After reviewing these tables, it should be easy to follow the Fortran programs for the algorithms. The standard Fortran library does not include a bitcount function, but it does include a function $IBITS(I, POS, LEN)$, which gives

Table 5.1   Recursive subdivision of a triangle $(p = 2)$.

k	$b(k)$	$(k_1, k_2)^{(0)}$	$(k_1, k_2)^{(1)}$	$(k_1, k_2)^{(2)}$	$\mathbf{x}^{(0)}$	$\mathbf{x}^{(1)}$	$\mathbf{x}^{(2)}$
$0 = 00$	0	$(0,0)$	$(0,1)$	$(0,2)$	$\mathbf{x}_0$	$\frac{1}{2}(\mathbf{x}_0 + \mathbf{x}_1)$	$\frac{1}{2}(\mathbf{x}_0 + \mathbf{x}_2)$
$1 = 01$	1	$(1,1)$	$(0,1)$	$(0,2)$	$\mathbf{x}_1$	$\frac{1}{2}(\mathbf{x}_0 + \mathbf{x}_1)$	$\frac{1}{2}(\mathbf{x}_0 + \mathbf{x}_2)$
$2 = 10$	1	$(1,1)$	$(1,2)$	$(0,2)$	$\mathbf{x}_1$	$\frac{1}{2}(\mathbf{x}_1 + \mathbf{x}_2)$	$\frac{1}{2}(\mathbf{x}_0 + \mathbf{x}_2)$
$3 = 11$	2	$(2,2)$	$(1,2)$	$(0,2)$	$\mathbf{x}_2$	$\frac{1}{2}(\mathbf{x}_1 + \mathbf{x}_2)$	$\frac{1}{2}(\mathbf{x}_0 + \mathbf{x}_2)$

Table 5.2   Symmetric subdivision of a triangle $(p = 2)$.

k	$b(k)$	$(k_1, k_2)^{(0)}$	$(k_1, k_2)^{(1)}$	$(k_1, k_2)^{(2)}$	$\mathbf{x}^{(0)}$	$\mathbf{x}^{(1)}$	$\mathbf{x}^{(2)}$
$0 = 00$	0	$(0,0)$	$(0,1)$	$(0,2)$	$\mathbf{x}_0$	$\frac{1}{2}(\mathbf{x}_0 + \mathbf{x}_1)$	$\frac{1}{2}(\mathbf{x}_0 + \mathbf{x}_2)$
$1 = 01$	1	$(0,1)$	$(1,1)$	$(1,2)$	$\frac{1}{2}(\mathbf{x}_0 + \mathbf{x}_1)$	$\mathbf{x}_1$	$\frac{1}{2}(\mathbf{x}_1 + \mathbf{x}_2)$
$2 = 10$	1	$(0,1)$	$(0,2)$	$(1,2)$	$\frac{1}{2}(\mathbf{x}_0 + \mathbf{x}_1)$	$\frac{1}{2}(\mathbf{x}_0 + \mathbf{x}_2)$	$\frac{1}{2}(\mathbf{x}_1 + \mathbf{x}_2)$
$3 = 11$	2	$(0,2)$	$(1,2)$	$(2,2)$	$\frac{1}{2}(\mathbf{x}_0 + \mathbf{x}_2)$	$\frac{1}{2}(\mathbf{x}_1 + \mathbf{x}_2)$	$\mathbf{x}_2$

the value of the substring of bits of length LEN, starting at position POS, of the argument I. Thus, $IBITS(k, j, 1)$ gives the binary digit (0 or 1) at position j in the binary representation of k, which is all the information needed for the algorithm.

The Fortran subroutines for the recursive and symmetric subdivision algorithms are as follows. In these programs, ip denotes the dimension p, $klim$ denotes $2^p - 1$, $ivertold$ is the array of vertices of the original simplex, $ivold$ is the array of vectors denoted by $\mathbf{V}(k_1, k_2)$ in the discussion above, and $ivnew$ is the array of vertices of the 2^p subdivided simplexes.

```
        subroutine symmetric(ivnew,ivertold,ip,klim)
c symmetric simplex subdivision; assumes ip<32
        implicit integer(i-l)
        integer(2) ivold,ivnew,ivertold
        dimension ivold(1:ip,0:ip,0:ip),ivnew(1:ip,0:klim,0:ip)
        dimension ivertold(1:ip,0:ip)
        ivold(1:ip,0:ip,0:ip)=0
        do 5 j=0,ip
        do 5 i=0,j
        ivold(1:ip,i,j)=ivertold(1:ip,i)/2+ivertold(1:ip,j)/2
```

```
 5 continue
      do 10 k=0,klim
      ibitct=0
      do 20 j=0,31
      ibitct=ibitct+ibits(k,j,1)
20 continue
      ivnew(1:ip,k,0)=ivold(1:ip,0,ibitct)
      k1=0
      k2=ibitct
      do 30 ll=1,ip
      if(ibits(k,ll-1,1).eq.0) then
            k2=k2+1
      else if(ibits(k,ll-1,1).eq.1) then
            k1=k1+1
      end if
      ivnew(1:ip,k,ll)=ivold(1:ip,k1,k2)
30 continue
10 continue
      return
      end


      subroutine recursive(ivnew,ivertold,ip,klim)
c recursive simplex subdivision; assumes ip<32
      implicit integer(i-l)
      integer(2) ivold,ivnew,ivertold
      dimension ivold(1:ip,0:ip,0:ip),ivnew(1:ip,0:klim,0:ip)
      dimension ivertold(1:ip,0:ip)
      ivold(1:ip,0:ip,0:ip)=0
      do 5 j=0,ip
      do 5 i=0,j
      ivold(1:ip,i,j)=ivertold(1:ip,i)/2+ivertold(1:ip,j)/2
 5 continue
      do 10 k=0,klim
      ibitct=0
      do 20 j=0,31
      ibitct=ibitct+ibits(k,j,1)
20 continue
      ivnew(1:ip,k,0)=ivold(1:ip,ibitct,ibitct)
      k1=ibitct
      k2=ibitct
      do 30 ll=1,ip
```

```
      if(ibits(k,ll−1,1).eq.0) then
            k2=k2+1
      else if(ibits(k,ll−1,1).eq.1) then
            k1=k1−1
      end if
      ivnew(1:ip,k,ll)=ivold(1:ip,k1,k2)
30 continue
10 continue
      return
      end
```

5.2.2 *Simplex subdivision properties*

Moore's simplex subdivision algorithms have a number of properties that will be useful in applying them to p-dimensional integration.

(1) As noted by Moore, the subdivided simplexes all have equal volume, equal to the initial simplex volume divided by 2^p. This follows from the fact that the general formula for the volume of a simplex with vertices $\mathbf{x}_0, \mathbf{x}_1, \ldots, \mathbf{x}_p$ is given in terms of $\Delta\mathbf{x}_a \equiv \mathbf{x}_a - \mathbf{x}_0$ by

$$V = \frac{1}{p!}|\det(\Delta x_{ai})|, \tag{5.12}$$

with rows of the determinant labeled by the vertex index $a = 1, \ldots, p$ and with columns of the determinant labeled by the vector component index $i = 1, \ldots, p$. Applying this to the vertices for the subdivided simplexes in Tables 5.1–5.4 verifies this statement for $p = 2, 3$, while a proof in the general case is given in Edelsbrunner and Grayson (2000).

(2) Again as noted by Moore, both the recursive and symmetric algorithms subdivide Kuhn simplexes into Kuhn simplexes, which however do not all have the same orientation, as illustrated in Figs. 5.3 and 5.4. This follows from the fact that Kuhn simplexes are a tiling of hypercubes, which are divided into hypercubes by axis parallel planes that intersect the midpoints of the hypercube edges. Adding additional diagonal slices intersecting the midpoints of the hypercube edges gives Kuhn tilings

Table 5.3   Recursive subdivision of a tetrahedron ($p = 3$).

k	$b(k)$	$(k_1,k_2)^{(0)}$	$(k_1,k_2)^{(1)}$	$(k_1,k_2)^{(2)}$	$(k_1,k_2)^{(3)}$	$x^{(0)}$	$x^{(1)}$	$x^{(2)}$	$x^{(3)}$
$0 = 000$	0	$(0,0)$	$(0,1)$	$(0,2)$	$(0,3)$	x_0	$\frac{1}{2}(x_0+x_1)$	$\frac{1}{2}(x_0+x_2)$	$\frac{1}{2}(x_0+x_3)$
$1 = 001$	1	$(1,1)$	$(0,1)$	$(0,2)$	$(0,3)$	x_1	$\frac{1}{2}(x_0+x_1)$	$\frac{1}{2}(x_0+x_2)$	$\frac{1}{2}(x_0+x_3)$
$2 = 010$	1	$(1,1)$	$(1,2)$	$(0,2)$	$(0,3)$	x_1	$\frac{1}{2}(x_1+x_2)$	$\frac{1}{2}(x_0+x_2)$	$\frac{1}{2}(x_0+x_3)$
$3 = 011$	2	$(2,2)$	$(1,2)$	$(0,2)$	$(0,3)$	x_2	$\frac{1}{2}(x_1+x_2)$	$\frac{1}{2}(x_0+x_2)$	$\frac{1}{2}(x_0+x_3)$
$4 = 100$	1	$(1,1)$	$(1,2)$	$(1,3)$	$(0,3)$	x_1	$\frac{1}{2}(x_1+x_2)$	$\frac{1}{2}(x_1+x_3)$	$\frac{1}{2}(x_0+x_3)$
$5 = 101$	2	$(2,2)$	$(1,2)$	$(1,3)$	$(0,3)$	x_2	$\frac{1}{2}(x_1+x_2)$	$\frac{1}{2}(x_1+x_3)$	$\frac{1}{2}(x_0+x_3)$
$6 = 110$	2	$(2,2)$	$(2,3)$	$(1,3)$	$(0,3)$	x_2	$\frac{1}{2}(x_2+x_3)$	$\frac{1}{2}(x_1+x_3)$	$\frac{1}{2}(x_0+x_3)$
$7 = 111$	3	$(3,3)$	$(2,3)$	$(1,3)$	$(0,3)$	x_3	$\frac{1}{2}(x_2+x_3)$	$\frac{1}{2}(x_1+x_3)$	$\frac{1}{2}(x_0+x_3)$

Table 5.4   Symmetric subdivision of a tetrahedron ($p = 3$).

k	$b(k)$	$(k_1,k_2)^{(0)}$	$(k_1,k_2)^{(1)}$	$(k_1,k_2)^{(2)}$	$(k_1,k_2)^{(3)}$	$\mathbf{x}^{(0)}$	$\mathbf{x}^{(1)}$	$\mathbf{x}^{(2)}$	$\mathbf{x}^{(3)}$
$0 = 000$	0	$(0,0)$	$(0,1)$	$(0,2)$	$(0,3)$	$\mathbf{x}_0$	$\frac{1}{2}(\mathbf{x}_0+\mathbf{x}_1)$	$\frac{1}{2}(\mathbf{x}_0+\mathbf{x}_2)$	$\frac{1}{2}(\mathbf{x}_0+\mathbf{x}_3)$
$1 = 001$	1	$(0,1)$	$(1,1)$	$(1,2)$	$(1,3)$	$\frac{1}{2}(\mathbf{x}_0+\mathbf{x}_1)$	$\mathbf{x}_1$	$\frac{1}{2}(\mathbf{x}_1+\mathbf{x}_2)$	$\frac{1}{2}(\mathbf{x}_1+\mathbf{x}_3)$
$2 = 010$	1	$(0,1)$	$(0,2)$	$(1,2)$	$(1,3)$	$\frac{1}{2}(\mathbf{x}_0+\mathbf{x}_1)$	$\frac{1}{2}(\mathbf{x}_0+\mathbf{x}_2)$	$\frac{1}{2}(\mathbf{x}_1+\mathbf{x}_2)$	$\frac{1}{2}(\mathbf{x}_1+\mathbf{x}_3)$
$3 = 011$	2	$(0,2)$	$(1,2)$	$(2,2)$	$(2,3)$	$\frac{1}{2}(\mathbf{x}_0+\mathbf{x}_2)$	$\frac{1}{2}(\mathbf{x}_1+\mathbf{x}_2)$	$\mathbf{x}_2$	$\frac{1}{2}(\mathbf{x}_2+\mathbf{x}_3)$
$4 = 100$	1	$(0,1)$	$(0,2)$	$(0,3)$	$(1,3)$	$\frac{1}{2}(\mathbf{x}_0+\mathbf{x}_1)$	$\frac{1}{2}(\mathbf{x}_0+\mathbf{x}_2)$	$\frac{1}{2}(\mathbf{x}_0+\mathbf{x}_3)$	$\frac{1}{2}(\mathbf{x}_1+\mathbf{x}_3)$
$5 = 101$	2	$(0,2)$	$(1,2)$	$(1,3)$	$(2,3)$	$\frac{1}{2}(\mathbf{x}_0+\mathbf{x}_2)$	$\frac{1}{2}(\mathbf{x}_1+\mathbf{x}_2)$	$\frac{1}{2}(\mathbf{x}_1+\mathbf{x}_3)$	$\frac{1}{2}(\mathbf{x}_2+\mathbf{x}_3)$
$6 = 110$	2	$(0,2)$	$(0,3)$	$(1,3)$	$(2,3)$	$\frac{1}{2}(\mathbf{x}_0+\mathbf{x}_2)$	$\frac{1}{2}(\mathbf{x}_0+\mathbf{x}_3)$	$\frac{1}{2}(\mathbf{x}_1+\mathbf{x}_3)$	$\frac{1}{2}(\mathbf{x}_2+\mathbf{x}_3)$
$7 = 111$	3	$(0,3)$	$(1,3)$	$(2,3)$	$(3,3)$	$\frac{1}{2}(\mathbf{x}_0+\mathbf{x}_3)$	$\frac{1}{2}(\mathbf{x}_1+\mathbf{x}_3)$	$\frac{1}{2}(\mathbf{x}_2+\mathbf{x}_3)$	$\mathbf{x}_3$

of both the original and the subdivided hypercubes. However, as also noted by Moore, when the algorithms are applied to general simplexes, the resultant subdivided simplexes can have different shapes, and are not isomorphic. For $p = 2$, Fig. 5.1 shows that recursive subdivision applied to the standard simplex leads to subsimplexes of different shapes, while Fig. 5.2 shows that symmetric subdivision applied to the standard simplex leads to subsimplexes that are all standard simplexes with dimension reduced by half. However, an examination of the vertices in Table 5.4 shows that already at $p = 3$, symmetric subdivision of a standard simplex does not lead to subsimplexes that are all half size standard simplexes. For example, for $k = 2$ in Table 5.4, there are vertices $\frac{1}{2}(\mathbf{x}_0 + \mathbf{x}_2)$ and $\frac{1}{2}(\mathbf{x}_1 + \mathbf{x}_3)$, the edge joining which has length $\sqrt{3}/2$, whereas the maximum side length of a half size $p = 3$ standard simplex is $\sqrt{2}/2$.

(3) An important question is whether the maximum side length of the subdivided simplexes decreases at each stage of subdivision. For Kuhn simplexes, the answer is immediate, since subdivision results in Kuhn simplexes of half the dimension. Since the longest side of a unit Kuhn simplex in dimension p has length $\sqrt{p}$, after ℓ subdivisions the maximum side length will be

$$L_{\max}^{\text{Kuhn}} = \sqrt{p}/2^{\ell}, \tag{5.13}$$

irrespective of whether recursive or symmetric subdivision is used. For standard simplexes, we can get an upper bound on the maximum side length by noting that a unit standard simplex on axes labeled by components $y_1, \ldots, y_p$ is obtained from a unit Kuhn simplex on axes labeled by components $x_1, \ldots, x_p$ by the linear transformation $y_p = x_p$, $y_{p-1} = x_{p-1} - x_p$, $y_{p-2} = x_{p-2} - x_{p-1}, \ldots, y_2 = x_2 - x_3$, $y_1 = x_1 - x_2$, since this maps the components of the Kuhn simplex vertices given in Eq. (2.5) to the corresponding components of the standard simplex vertices given in Eq. (2.3). By linearity, this relation also holds between vertices of corresponding subdivided simplexes obtained from the initial unit standard and Kuhn simplexes by applying the same midpoint subdivision method (either symmetric or recursive) successively to each. Consequently, the length L^{standard}

of an edge with components $E^{\mathrm{S}}_{1,\ldots,p}$ of a subdivided standard simplex can be expressed in terms of the components $E^{\mathrm{K}}_{1,\ldots,p}$ of the corresponding edge of the related Kuhn simplex by

$$L^{\mathrm{standard}} \equiv \left[\sum_{i=1}^{p}(E^{S}_{i})^2\right]^{\frac{1}{2}} = \left[\sum_{i=1}^{p-1}(E^{K}_{i} - E^{K}_{i+1})^2 + (E^{K}_{p})^2\right]^{\frac{1}{2}}$$

$$\leq 2\left[\sum_{i=1}^{p}(E^{K}_{i})^2\right]^{\frac{1}{2}} = 2L^{\mathrm{Kuhn}} . \tag{5.14}$$

Thus the length L^{standard} is bounded from above by twice the maximum length corresponding to a subdivided Kuhn simplex, and so

$$L^{\mathrm{standard}}_{\mathrm{max}} \leq \sqrt{p}/2^{\ell-1} . \tag{5.15}$$

We have verified this inequality numerically for both the recursive and symmetric subdivision algorithms. The numerical results suggest that the symmetric subdivision algorithm is in fact a factor of 2 better than the bound of Eq. (5.15), so that

$$L^{\mathrm{standard;\ symmetric}}_{\mathrm{max}} \leq \sqrt{p}/2^{\ell} , \tag{5.16}$$

but we do not have a proof of this. We already see evidence of this difference between the symmetric and recursive algorithms in Tables 5.3 and 5.4. As noted above, from Table 5.4 we saw that symmetric subdivision of a $p = 3$ standard simplex gives an edge of length $\sqrt{3}/2$, and it is easy to see that this is the longest edge. However, from Table 5.3 for recursive subdivision, we see that for $k = 5$ there are vertices $\mathbf{x}_2$ and $\frac{1}{2}(\mathbf{x}_1 + \mathbf{x}_3)$, the edge joining which, for an initial standard simplex, has length $\sqrt{6}/2$.

(4) The result of Eqs. (5.15) and (5.16) suggests the stronger conjecture, that after any number ℓ of symmetric (recursive) subdivisions of a standard simplex, the resulting subsimplexes each fit within a hypercube of side $1/2^{\ell}$ ($1/2^{\ell-1}$). A simple argument shows this to be true for $\ell = 1$ in any dimension p. Although we do not have a proof for general ℓ, we will use this conjecture in certain of the algorithms constructed below. For Kuhn simplexes, an analogous statement with a hypercube of side $1/2^{\ell}$ is true for both symmetric and recursive subdivision, as noted above in the discussion preceding Eq. (5.13).

(5) Finally, we note that although the symmetric algorithm gives the same simplex subdivision after permutation of the starting vertices in dimension $p = 2$, as can be verified from Table 5.2, it is not permutation symmetric in dimension $p = 3$, as can be verified from Table 5.4. For example, interchanging the labels 0 and 1 in the $k = 2$ line of Table 5.4 gives a set of vertices that is not in the table. This means that with symmetric (as well as recursive) subdivision in dimension $p \geq 3$, inequivalent simplex subdivisions can be generated by permuting the labels of the starting vertices. However, we have not incorporated this feature into our programs.

The properties just listed show that the symmetric and recursive subdivision algorithms are well suited to adaptive higher dimensional integration. They are easily computable in terms of the vertex coordinates for a general simplex, and give subsimplexes of equal volume, so that it is not necessary to calculate a determinant to obtain the volume. Additionally, the bound on the maximum side length decreases as a constant times $1/2^{\ell}$ with increasing order of subdivision ℓ, so that for highly differentiable functions the application of high order integration formulas gives errors that decrease rapidly with ℓ.

5.3   Hypercube and hyper-rectangle subdivision

5.3.1   *Hypercube subdivision algorithm*

We have discussed simplexes first because our direct approach to hypercube integration will be based on following as closely as possible the methods that we develop for simplex integration. In our direct hypercube programs (i.e., the ones not based on tiling a side 1 hypercube with Kuhn simplexes), we will start from a half-side 1 hypercube with base region

$$(-1,1) \otimes (-1,1) \otimes \cdots \otimes (-1,1). \tag{5.17}$$

This region has inversion symmetry around the origin, and consequently the only monomials that have non-vanishing integrals over this region are ones in which *each* coordinate appears with an even exponent, considerably simplifying the calculations needed to construct higher order integration rules.

Since we restrict ourselves to axis-parallel hypercubes, only $p+1$ real numbers are needed to uniquely specify a hypercube: the p coordinates of the centroid $\mathbf{x}_c$ and the half-side length S. For example, for the region of Eq. (5.17), the centroid is $\mathbf{x}_c = (0, 0, \ldots, 0)$ and the half-side is 1. Once we have adopted this labeling, we can give a very simple subdivision algorithm for hypercubes, constructed in direct analogy with Moore's simplex subdivision algorithms.

The hypercube subdivision algorithm proceeds as follows. Start from a hypercube with centroid $\mathbf{x}_c$ and half-side S, with sides parallel to the p unit axis vectors

$$\hat{u}_1 = (1, 0, 0, \ldots, 0)$$
$$\hat{u}_2 = (0, 1, 0, \ldots, 0)$$
$$\cdots\cdots\cdots\cdots \tag{5.18}$$
$$\hat{u}_{p-1} = (0, 0, \ldots, 1, 0)$$
$$\hat{u}_p = (0, 0, \ldots, 0, 1)\,.$$

To subdivide it into 2^p subhypercubes, take the new half-side as $S/2$. To get the new centroids $\mathbf{x}_{c;k}$, labeled by $k = 0, \ldots, 2^p - 1$, scan along the binary representation of k from right (the units digit) to left. Denoting the p digits in this representation by $1 \leq j \leq p$, let us label the units digit as $j = 1$, the power of 2 digit as $j = 2$, the power of 4 digit as $j = 3$, and so forth. For all $1 \leq j \leq p$, if the jth digit is 0, add $\frac{1}{2}S\hat{u}_j$ to $\mathbf{x}_c$, and if the jth digit is 1, add $-\frac{1}{2}S\hat{u}_j$ to $\mathbf{x}_c$. For each given k, this gives the centroid of the kth subhypercube. This algorithm is illustrated for the case of a cube ($p = 3$) in Table 5.5. This algorithm is simpler than the ones for subdividing simplexes, since it only needs the Fortran IBITS function, but does not require subsequent computation of the bitcount function.

The Fortran subroutine for hypercube subdivision follows, again with *ip* the dimension p and *klim* $= 2^p - 1$. The array *icenterold* stores the half-side of the original hypercube in the index 0 location, and the centroid coordinates in the index locations $1, \ldots, p$. The array *icenternew* stores the half-sides and centroid coordinates of the 2^p subdivided hypercubes, using the same storage convention.

Table 5.5   Subdivision of a cube of half-side S and centroid $\mathbf{x}_c$ ($p = 3$).

k	$\mathbf{x}_{c;k} - \mathbf{x}_c$
$0 = 000$	$(S/2, S/2, S/2)$
$1 = 001$	$(-S/2, S/2, S/2)$
$2 = 010$	$(S/2, -S/2, S/2)$
$3 = 011$	$(-S/2, -S/2, S/2)$
$4 = 100$	$(S/2, S/2, -S/2)$
$5 = 101$	$(-S/2, S/2, -S/2)$
$6 = 110$	$(S/2, -S/2, -S/2)$
$7 = 111$	$(-S/2, -S/2, -S/2)$

```
      subroutine cubediv(icenternew,icenterold,ip,klim)
c hypercube subdivision
      implicit integer(i-l)
      integer(2) icenternew, isidenew,icenterold, isideold
      dimension icenterold(0:ip),icenternew(0:klim,0:ip)
      isideold=icenterold(0)
      isidenew=isideold/2
      icenternew(0:klim,0)=isidenew
      do 10 k=0,klim
      do 30 ll=1,ip
      if(ibits(k,ll-1,1).eq.0) then
          icenternew(k,ll)=icenterold(ll)+isidenew
      else if(ibits(k,ll-1,1).eq.1) then
          icenternew(k,ll)=icenterold(ll)-isidenew
      end if
30 continue
10 continue
      return
      end
```

5.3.2   *Hypercube and subdivision properties*

The hypercube subdivision algorithm evidently has properties analogous to those of the simplex subdivision algorithms: each subhypercube has the same volume, equal to the original hypercube volume divided by 2^p, and every linear dimension of each subhypercube is a factor of 2 smaller than the corresponding linear dimension of the hypercube that preceded it in the subdivision chain. This latter implies that after ℓ subdivisions, the resulting subhypercubes all have dimension reduced by a factor $1/2^\ell$.

For a hypercube with centroid $\mathbf{x}_c$ and half-side S, and for a general point $\mathbf{x}$, let us define the coordinate relative to the centroid as $\tilde{\mathbf{x}} = \mathbf{x} - \mathbf{x}_c$, as we did in the simplex case in Eq. (5.4). Consider now the set of $2p$ points $\tilde{\mathbf{x}}_a$, $a = 1, \ldots, 2p$ defined by

$$
\begin{aligned}
\tilde{\mathbf{x}}_1 &= (S, 0, 0, \ldots, 0) \\
\tilde{\mathbf{x}}_2 &= (0, S, 0, \ldots, 0) \\
&\quad \cdots\cdots\cdots\cdots \\
\tilde{\mathbf{x}}_p &= (0, 0, \ldots, S) \\
\tilde{\mathbf{x}}_{p+1} &= (-S, 0, 0, \ldots, 0) \\
\tilde{\mathbf{x}}_{p+2} &= (0, -S, 0, \ldots, 0) \\
&\quad \cdots\cdots\cdots\cdots \\
\tilde{\mathbf{x}}_{2p} &= (0, 0, \ldots, -S) \, .
\end{aligned}
\tag{5.19}
$$

These points are the centroids of the maximal boundary hypercubes, and will play a role in the direct hypercube algorithm analogous to that played by the simplex vertices in the simplex adaptive algorithm. For future use, we need the following result, analogous to that of Eqs. (5.9) and (5.10) for the simplex case. Consider the sum

$$
\tilde{\mathbf{X}} = \sum_{a=1}^{N} \lambda_a \tilde{\mathbf{x}}_a \, ,
\tag{5.20}
$$

with the coefficients λ_a obeying

$$
\lambda_a > 0, \quad a = 1, \ldots, N \, , \quad \sum_{a=1}^{N} \lambda_a < 1 \, ,
\tag{5.21}
$$

with the points $\tilde{\mathbf{x}}_a$ any of the hypercube boundary points of Eq. (5.19). Some of these points may be omitted (in which case the corresponding coefficient λ_a is 0), and some used more than once, in the sum of Eq. (5.20). Then the point $\tilde{\mathbf{X}}$ lies inside the hypercube. To see this, we note that the projection of $\tilde{\mathbf{X}}$ along any axis i is of the form $\tilde{X}_i = S(\mu_+ - \mu_-)$, with $\mu_\pm$ each a sum of some subset of the coefficients λ_a, and hence $0 \le \mu_\pm < 1$. Therefore $-S < -S\mu_- \le \tilde{X}_i \le S\mu_+ < S$ for each axis component $\tilde{X}_i$, and thus $\mathbf{X}$ lies within the hypercube. This proof, again, is simpler than the corresponding result in the simplex case.

5.3.3 *Hyper-rectangle subdivision algorithm*

A simple variant of the algorithm used to subdivide hypercubes is used to subdivide hyper-rectangles along a specified set of axes i which have $iflag(i) = 1$. As before, ip is the dimension p, but now $klim1 = 2^{\sum_i iflag(i)} - 1$, so that the subdivided hyper-rectangles are indexed by $0, \ldots, klim1$. There are now p half-sides, so the input and output arrays are larger. The array *icenterold* stores the half-sides of the original hyper-rectangle in the index locations 1 to p, and the centroid coordinates in the index locations $p + 1, \ldots, 2p$. The array *icenternew* stores the half-sides and centroid coordinates of the $klim1 + 1$ subdivided hyper-rectangles, using the same storage convention. Only axes with $iflag = 1$ are halved, so for axes with $iflag = 0$ the original half-side and centroid coordinate are copied to the output array, while when $iflag = 1$, a binary logic similar to that used above for hypercubes is applied to calculate the new half-sides and centroid coordinates.

```
      subroutine hyperrectdiv(ivnew,ivertold,iflag,ip,klim1)
c hyperrectangle subdivision; iflag=1 for directions to be halved
c sides stored 1:ip, center coordinates stored ip+1:2*ip
      implicit integer(i-l)
      integer(2) ivnew,ivertold,isidell
      dimension ivertold(1:2*ip),ivnew(0:klim1,1:2*ip)
      dimension iflag(1:ip)
      do 10 k=0,klim1
      llsum=0
      do 30 ll=1,ip
      llsum=llsum+iflag(ll)
```

```
      if(iflag(ll).eq.0) then
            ivnew(k,ll)=ivertold(ll)
            ivnew(k,ip+ll)=ivertold(ip+ll)
      else if(iflag(ll).eq.1) then
            isidell=ivertold(ll)/2
            ivnew(k,ll)=isidell
            if(ibits(k,llsum−1,1).eq.0) then
                  ivnew(k,ip+ll)=ivertold(ip+ll)+isidell
            else if(ibits(k,llsum−1,1).eq.1) then
                  ivnew(k,ip+ll)=ivertold(ip+ll)−isidell
            end if
      end if
30 continue
10 continue
      return
      end
```

5.4 Vandermonde solvers

Since we will repeatedly encounter Vandermonde equations in setting up parameterized higher order integration formulas, both for simplexes and for hypercubes, we digress at this point to discuss methods of solving a Vandermonde system. We write an order N Vandermonde system in the standard form

$$\sum_{j=1}^{N} y_j^{k-1} w_j = q_k, \quad k = 1, \ldots, N \ .$$

(5.22)

The explicit inversion of the Vandermonde system is well known (see, e.g., Neagoe (1996), Heinen and Niederjohn (1997)), and takes the form

$w_1 =$

$$\frac{q_N - S_1(y_2, \ldots, y_N)q_{N-1} + S_2(y_2, \ldots, y_N)q_{N-2} - \cdots + (-1)^{N-1} y_2 \cdots y_N q_1}{(y_1 - y_2)(y_1 - y_3) \cdots (y_1 - y_N)},$$

(5.23)

with $S_j(y_2, \ldots, y_N)$ the sum of j-fold products of $y_2, \ldots, y_N$,

$$S_1(y_2, \ldots, y_N) = y_2 + \cdots + y_N \,,$$
$$S_2(y_2, \ldots, y_N) = y_2 y_3 + \cdots + y_2 y_N + y_3 y_4 + \cdots + y_3 y_N$$
$$+ \cdots + y_{N-1} y_N \,, \qquad (5.24)$$

and so forth. The remaining unknowns w_2 through w_N are obtained from this formula by cyclic permutation of the indices $j = 1, \ldots, N$ on the w_j and the y_j, with the q_k held fixed. For N not too large it is straightforward to program this solution, and we include subroutines for the $N = 2, 3, 4, 6, 8$ cases in the programs. This suffices to solve the Vandermonde equations appearing in the fifth through ninth order simplex formulas, and in the fifth through ninth order hypercube formulas derived below.

For large N, programming the explicit solution becomes inefficient and a better procedure is to use a compact algorithm for solving the Vandermonde equations for general N, based on polynomial operations, which has running time proportional to N^2. A good method of this type is the algorithm vander.for given in the book *Numerical Recipes in Fortran 77* by Press et al. (1992).

5.5 Parameterized higher order integration formulas for a general simplex

We turn next to deriving higher order integration formulas for a general simplex, which are expressed directly in terms of the set of simplex vertices, and which involve parameters that can be changed to sample the function over the simplex in different ways. Two different choices of the parameters then give two different integration rules of the same order, which can be compared to give a local error estimate for use in adaptive integration.

5.5.1 *Simplex integrals in terms of moments*

Since we want to derive integration rules up to ninth order in accuracy, we start from an expansion of a general function $f(\tilde{\mathbf{x}})$ up to ninth order, with $\tilde{\mathbf{x}}$ as before the p-dimensional coordinate referred to the simplex centroid as origin. The expansion reads,

$$
\begin{aligned}
f(\tilde{\mathbf{x}}) = {} & A + B_{i_1}\tilde{x}_{i_1} + C_{i_1 i_2}\tilde{x}_{i_1}\tilde{x}_{i_2} + D_{i_1 i_2 i_3}\tilde{x}_{i_1}\tilde{x}_{i_2}\tilde{x}_{i_3} \\
& + E_{i_1 i_2 i_3 i_4}\tilde{x}_{i_1}\tilde{x}_{i_2}\tilde{x}_{i_3}\tilde{x}_{i_4} + F_{i_1\cdots i_5}\tilde{x}_{i_1}\cdots\tilde{x}_{i_5} \\
& + G_{i_1\cdots i_6}\tilde{x}_{i_1}\cdots\tilde{x}_{i_6} + H_{i_1\cdots i_7}\tilde{x}_{i_1}\cdots\tilde{x}_{i_7} \\
& + I_{i_1\cdots i_8}\tilde{x}_{i_1}\cdots\tilde{x}_{i_8} + J_{i_1\cdots i_9}\tilde{x}_{i_1}\cdots\tilde{x}_{i_9} + \cdots ,
\end{aligned}
\tag{5.25}
$$

with repeated component indices summed 1 to p. We next need expressions for the integral of the monomials appearing in the expansion of Eq. (5.25) over a general simplex with vertices $\mathbf{x}_0, \ldots, \mathbf{x}_p$. A general formula for these integrals has been given by Good and Gaskins (1969, 1971). They define $m(\nu)$ as the generalized moment

$$
m(\nu) = \int_{\text{simplex}} dx_1 \cdots dx_p \, \tilde{x}_1^{\nu_1} \cdots \tilde{x}_p^{\nu_p} ,
\tag{5.26}
$$

and show that $m(\nu)$ is equal to the coefficient of $t_1^{\nu_1} \cdots t_p^{\nu_p}$ in the expansion of

$$
\frac{Vp!\,\nu_1!\cdots\nu_p!}{(p+\nu_1+\cdots+\nu_p)!} \exp\left[\sum_{s=2}^{\infty} \frac{1}{s} W_s\right] .
\tag{5.27}
$$

Here W_s is a double sum over ith components of the $p+1$ simplex vertices labeled by $a = 0, \ldots, p$, given by

$$
W_s = \sum_{a=0}^{p}\left[\sum_{i=1}^{p} \tilde{x}_{ai} t_i\right]^s ,
\tag{5.28}
$$

and V is the simplex volume. Good and Gaskins derive this formula by first transforming the original simplex to a standard simplex, followed by lengthy algebraic manipulations to express the resulting formula symmetrically in terms of standard simplex vertices. We give in Appendix B a derivation that proceeds directly, and with manifest symmetry, from the vertices of the original simplex.

To proceed to 9th order we need an expansion of the exponential in Eq. (5.27) through 9th order. Since each W_s is of degree s in the coordinates, the terms in this expansion are as follows:

second order : $\dfrac{W_2}{2}$

third order : $\dfrac{W_3}{3}$

fourth order : $\dfrac{W_2^2}{8} + \dfrac{W_4}{4}$

fifth order : $\dfrac{W_2 W_3}{6} + \dfrac{W_5}{5}$

sixth order : $\dfrac{W_2^3}{48} + \dfrac{W_3^2}{18} + \dfrac{W_2 W_4}{8} + \dfrac{W_6}{6}$

seventh order : $\dfrac{W_2^2 W_3}{24} + \dfrac{W_3 W_4}{12} + \dfrac{W_2 W_5}{10} + \dfrac{W_7}{7}$

eighth order : $\dfrac{W_2^4}{384} + \dfrac{W_2 W_3^2}{36} + \dfrac{W_2^2 W_4}{32} + \dfrac{W_4^2}{32} + \dfrac{W_3 W_5}{15} + \dfrac{W_2 W_6}{12}$

$\qquad\qquad + \dfrac{W_8}{8}$

ninth order : $\dfrac{W_2^3 W_3}{144} + \dfrac{W_3^3}{162} + \dfrac{W_2 W_3 W_4}{24} + \dfrac{W_2^2 W_5}{40} + \dfrac{W_4 W_5}{20}$

$$+ \dfrac{W_3 W_6}{18} + \dfrac{W_2 W_7}{14} + \dfrac{W_9}{9}. \qquad (5.29)$$

We are interested in integrals of monomials of the form $\tilde{x}_{i_1} \cdots \tilde{x}_{i_n}$, with $i_1, \ldots, i_n$ ranging from 1 to p. Good and Gaskins note that it suffices to consider the case in which all the indices $i_1, \ldots, i_n$ are distinct (which is always possible for $p \geq n$), since the combinatoric factors are such that this gives a result that also applies to the case when some of the component indices are equal, as must necessarily be the case when $p < n$. So we can take $\nu_i = 1$, $i = 1, \ldots, n$, and $\sum_i \nu_i = n$, with n the order of the monomial. We now can infer from Eq. (5.29) the moment integrals

$$\frac{1}{V} \int_{\text{simplex}} dx_1 \cdots dx_p \tilde{x}_{i_1} \cdots \tilde{x}_{i_n} = \frac{p!}{(p+n)!} \mathcal{S}_n, \qquad (5.30)$$

with the quantities $\mathcal{S}_n$ (with tensor indices suppressed) given in terms of tensors $S_{i_1\cdots i_n}$ defined by sums over the vertices,

$$S_{i_1\cdots i_n} = \sum_{a=0}^{p} \tilde{x}_{ai_1} \cdots \tilde{x}_{ai_n}, \qquad (5.31)$$

as follows:

$$\mathcal{S}_2 = S_{i_1 i_2}$$

$$\mathcal{S}_3 = 2 S_{i_1 i_2 i_3}$$

$$\mathcal{S}_4 = S_{i_1 i_2} S_{i_3 i_4} + S_{i_1 i_3} S_{i_2 i_4} + S_{i_1 i_4} S_{i_2 i_3} + 6 S_{i_1 i_2 i_3 i_4} = S_{i_1 i_2} S_{i_3 i_4}$$
$$+ \, 2\,\text{terms} + 6 S_{i_1 i_2 i_3 i_4}$$

$$\mathcal{S}_5 = 2(S_{i_1 i_2} S_{i_3 i_4 i_5} + 9\,\text{terms}) + 24 S_{i_1 i_2 i_3 i_4 i_5}$$

$$\mathcal{S}_6 = S_{i_1 i_2} S_{i_3 i_4} S_{i_5 i_6} + 14\,\text{terms} + 4(S_{i_1 i_2 i_3} S_{i_4 i_5 i_6} + 9\,\text{terms})$$
$$+ \, 6(S_{i_1 i_2} S_{i_3 i_4 i_5 i_6} + 14\,\text{terms}) + 120 S_{i_1 i_2 i_3 i_4 i_5 i_6}$$

$$\mathcal{S}_7 = 2(S_{i_1 i_2} S_{i_3 i_4} S_{i_5 i_6 i_7} + 104\,\text{terms}) + 12(S_{i_1 i_2 i_3} S_{i_4 i_5 i_6 i_7} + 34\,\text{terms})$$
$$+ \, 24(S_{i_1 i_2} S_{i_3 i_4 i_5 i_6 i_7} + 20\,\text{terms}) + 720 S_{i_1 i_2 i_3 i_4 i_5 i_6 i_7}$$

$$\mathcal{S}_8 = S_{i_1 i_2} S_{i_3 i_4} S_{i_5 i_6} S_{i_7 i_8} + 104\,\text{terms} + 4(S_{i_1 i_2} S_{i_3 i_4 i_5} S_{i_6 i_7 i_8} + 279\,\text{terms})$$
$$+ \, 6(S_{i_1 i_2} S_{i_3 i_4} S_{i_5 i_6 i_7 i_8} + 209\,\text{terms})$$
$$+ \, 36(S_{i_1 i_2 i_3 i_4} S_{i_5 i_6 i_7 i_8} + 34\,\text{terms}) + 48(S_{i_1 i_2 i_3} S_{i_4 i_5 i_6 i_7 i_8} + 55\,\text{terms})$$
$$+ \, 120(S_{i_1 i_2} S_{i_3 i_4 i_5 i_6 i_7 i_8} + 27\,\text{terms}) + 5040 S_{i_1 i_2 i_3 i_4 i_5 i_6 i_7 i_8}$$

$$\mathcal{S}_9 = 2(S_{i_1 i_2} S_{i_3 i_4} S_{i_5 i_6} S_{i_7 i_8 i_9} + 1259\,\text{terms}) + 8(S_{i_1 i_2 i_3} S_{i_4 i_5 i_6} S_{i_7 i_8 i_9}$$
$$+ \, 279\,\text{terms}) + 12(S_{i_1 i_2} S_{i_3 i_4 i_5} S_{i_6 i_7 i_8 i_9} + 1259\,\text{terms})$$
$$+ \, 24(S_{i_1 i_2} S_{i_3 i_4} S_{i_5 i_6 i_7 i_8 i_9} + 377\,\text{terms}) + 144(S_{i_1 i_2 i_3 i_4} S_{i_5 i_6 i_7 i_8 i_9}$$
$$+ \, 125\,\text{terms}) + 240(S_{i_1 i_2 i_3} S_{i_4 i_5 i_6 i_7 i_8 i_9} + 83\,\text{terms})$$
$$+ \, 720(S_{i_1 i_2} S_{i_3 i_4 i_5 i_6 i_7 i_8 i_9} + 35\,\text{terms}) + 40320 S_{i_1 i_2 i_3 i_4 i_5 i_6 i_7 i_8 i_9} \,.$$
$$(5.32)$$

The rule for forming terms in Eq. (5.32) from those in Eq. (5.29) is this: for each W_s in Eq. (5.29) there is a tensor factor S with s indices, and the product of such factors appears repeated in all nontrivial index permutations, giving the "terms" not shown explicitly in Eq. (5.32). The numerical coefficient is constructed from the denominator appearing in Eq. (5.29), multiplied by a numerator consisting of a factor $s!$ for each W_s, and for each W_s^m an additional factor $m!$

(that is, for W_s^m there is altogether a factor $(s!)^m m!$). For example, a W_2^3 in Eq. (5.29) gives rise to a numerator factor of $(2!)^3 3! = 48$ in Eq. (5.32), and a $W_2 W_3 W_4$ in Eq. (5.29) gives rise to a numerator factor of $2! 3! 4! = 288$ in Eq. (5.32). In each case, the product of this numerator factor, times the number of terms in the symmetrized expansion, is equal to $n!$. For example, $48 \times 15 = 720 = 6!$, and $288 \times 1260 = 362880 = 9!$.

Our next step is to combine Eqs. (5.25), (5.30), and (5.32) to get a formula for the integral of the function f over a general simplex, expressed in terms of its expansion coefficients. Since we will always be dealing with symmetrized tensors, it is useful at this point to condense the notation, by labelling the contractions of the expansion coefficients with the tensors S by the partition of n which appears. Thus, we will write

$$C_{i_1 i_2} S_{i_1 i_2} = C_2$$
$$F_{i_1 i_2 i_3 i_4 i_5}(S_{i_1 i_2} S_{i_3 i_4 i_5} + 9 \text{ terms}) = F_{3+2} \qquad (5.33)$$
$$H_{i_1 i_2 i_3 i_4 i_5 i_6 i_7}(S_{i_1 i_2} S_{i_3 i_4} S_{i_5 i_6 i_7} + 104 \text{ terms}) = H_{3+2+2} \,,$$

and so forth. Since the partitions of n that are relevant involve only integers ≥ 2, a complete list of partitions that appear through ninth order is as follows:

C 2

D 3

E $4,\ 2+2$

F $5,\ 3+2$

G $6,\ 4+2,\ 2+2+2,\ 3+3$

H $7,\ 5+2,\ 3+2+2,\ 4+3$

I $8,\ 6+2,\ 4+2+2,\ 2+2+2+2,\ 5+3,\ 3+3+2,\ 4+4$

J $9,\ 7+2,\ 5+2+2,\ 3+2+2+2,\ 4+3+2,\ 6+3,\ 5+4,$
$\qquad 3+3+3\,.$ $\qquad\qquad\qquad\qquad\qquad\qquad\qquad (5.34)$

Employing this condensed notation, we now get the following master formula for the integral of $f(\mathbf{x})$ over a general simplex,

$$\frac{1}{V} \int_{\text{simplex}} dx_1 \cdots dx_p f(\tilde{\mathbf{x}}) = A + \frac{p!}{(p+2)!} C_2 + \frac{p!}{(p+3)!} 2D_3$$

$$+ \frac{p!}{(p+4)!} (6E_4 + E_{2+2}) + \frac{p!}{(p+5)!} (24F_5 + 2F_{3+2})$$

$$+ \frac{p!}{(p+6)!} (120G_6 + 6G_{4+2} + 4G_{3+3} + G_{2+2+2})$$

$$+ \frac{p!}{(p+7)!} (720H_7 + 24H_{5+2} + 12H_{4+3} + 2H_{3+2+2})$$

$$+ \frac{p!}{(p+8)!} (5040I_8 + 120I_{6+2} + 48I_{5+3} + 36I_{4+4} + 6I_{4+2+2}$$

$$+ 4I_{3+3+2} + I_{2+2+2+2})$$

$$+ \frac{p!}{(p+9)!} (40320J_9 + 720J_{7+2} + 240J_{6+3} + 144J_{5+4}$$

$$+ 24J_{5+2+2} + 12J_{4+3+2} + 8J_{3+3+3} + 2J_{3+2+2+2})$$

$$+ \cdots . \tag{5.35}$$

Our procedure is now to match this expansion to discrete sums over the function f evaluated at points on the boundary or interior of the simplex. We will construct these sums using parameter multiples of the vertices of the simplex,

$$\Sigma_1(\lambda) = \sum_a f(\lambda \mathbf{x}_a), \quad 0 \leq \lambda \leq 1$$

$$\Sigma_2(\lambda, \sigma) = \sum_{a,b} f(\lambda \mathbf{x}_a + \sigma \mathbf{x}_b), \quad 0 \leq \lambda, \sigma, \quad \lambda + \sigma \leq 1,$$

$$\Sigma_3(\lambda, \sigma, \mu) = \sum_{a,b,c} f(\lambda \mathbf{x}_a + \sigma \mathbf{x}_b + \mu \mathbf{x}_c), \quad 0 \leq \lambda, \sigma, \mu, \tag{5.36}$$

$$\lambda + \sigma + \mu \leq 1,$$

$$\Sigma_4(\lambda, \sigma, \mu, \kappa) = \sum_{a,b,c,d} f(\lambda \mathbf{x}_a + \sigma \mathbf{x}_b + \mu \mathbf{x}_c + \kappa \mathbf{x}_d), \quad 0 \leq \lambda, \sigma, \mu, \kappa,$$

$$\lambda + \sigma + \mu + \kappa \leq 1,$$

where the conditions on the parameters $\lambda, \sigma, \mu, \kappa$ guarantee, by our discussion of simplex properties, that the points summed over do not lie outside the simplex. In Eq. (5.36) the summation limits for a, b,

c, d are 0 to p for simplexes, and will be 1 to $2p$ later on when we apply these formulas to hypercubes. Clearly, once we have a formula for Σ_4, we can get a formula for Σ_3 by setting $\kappa = 0$ and dividing by $p + 1$ (which becomes $2p$ in the hypercube case); we can then get Σ_2 by further setting $\mu = 0$ and dividing out another factor of $p + 1$, and so forth. In Appendix F we give the expansion of Σ_4 in terms of $f(\tilde{0}) = A$ and the contractions $C_2, \ldots, J_{3+2+2+2}$ appearing in Eq. (5.35).

Evidently, for simplexes Eq. (5.36) requires $(p + 1)^4$ function evaluations to compute Σ_4, $(p + 1)^3$ function evaluations to compute Σ_3, etc. For hypercubes, with the simplex vertices replaced by the $2p$ points of Eq. (5.19), Eq. (5.36) requires $(2p)^4$ function evaluations to compute Σ_4, $(2p)^3$ function evaluations to compute Σ_3, etc. Since it is known that the minimal number of function evaluations for a simplex integration method of order $2t + 1$ involves $p^t/t! + O(p^{t-1})$ function calls (Stroud (1971), Grundmann and Möller (1978)), and for a hypercube integration method of order $2t + 1$ involves $(2p)^t/t! + O(p^{t-1})$ function calls (Lyness, 1965), we will take the leading Σs in our integration formulas to have equal arguments, e.g., $\Sigma_4(\lambda, \lambda, \lambda, \lambda)$, $\Sigma_3(\lambda, \lambda, \lambda, \lambda)$, etc. This allows the parameterized integration formulas constructed below to have an optimal leading order power dependence on the space dimension p (but reflecting the parameter freedom, the non-leading power terms will not in general be minimal). In Appendix F we give formulas for evaluating Σs with repeated arguments using a minimum number of function calls.

With these preliminaries in hand, we are now ready to set up integration formulas of first through fourth, fifth, seventh, and ninth order, for integrals over general simplexes.

5.5.2   *First through third order simplex formulas*

We begin here with integration formulas of first through third order, which can be more useful than high order formulas for integrating functions that are not highly differentiable. Two different first order accurate estimates of the integral of Eq. (5.35) are clearly

$$I_a = f(\tilde{\mathbf{0}}) = A,$$
$$I_b = \Sigma_1(\lambda)/\xi = A + \text{second order},$$
(5.37)

with $\tilde{\mathbf{x}} = \tilde{\mathbf{0}}$ the simplex centroid, and $\xi = p + 1$. Evidently I_a is the dimension p analog of the dimension one center-of-bin rule, and when $\lambda = 1$, I_b is the dimension p analog of the dimension one trapezoidal rule.

To get a second order accurate formula, we have to match the terms

$$A + \frac{p!}{(p+2)!}C_2$$
(5.38)

in Eq. (5.35). Solving $\Sigma_1(\lambda) = \xi A + \lambda^2 C_2 + \cdots$ for C_2, we get

$$C_2 \simeq [\Sigma_1(\lambda) - \xi A]/\lambda^2,$$
(5.39)

which when substituted into Eq. (5.38) gives the second order accurate formula

$$I = \left(1 - \frac{p!}{(p+2)!}\frac{\xi}{\lambda^2}\right) f(\tilde{\mathbf{0}}) + \frac{p!}{(p+2)!}\frac{1}{\lambda^2}\Sigma_1(\lambda).$$
(5.40)

Using two different parameter values $\lambda_{a,b}$ gives two different second order accurate estimates $I_{a,b}$ of the integral.

We give two different methods of getting a third order accurate formula, both of which will play a role in the methods for getting higher odd order formulas. We first note that for $\lambda = 2/(p+3)$, we have

$$\Sigma_1(\lambda) = \xi A + \lambda^2[C_2 + 2D_3/(p+3)],$$
(5.41)

and so the coefficients of C_2 and D_3 are in the same ratio as appears in Eq. (5.35). Hence defining an overall multiplicative factor κ_1 to make both terms match in magnitude, and adding a multiple κ_0 of A to make this term match, we get a third order accurate formula

$$I_a = \kappa_1\Sigma_1(\lambda) + \kappa_0 f(\tilde{\mathbf{0}}),$$
$$\kappa_1 = \frac{p!}{(p+2)!}\lambda^{-2} = \frac{(p+3)^2}{4(p+2)(p+1)},$$
(5.42)
$$\kappa_0 = 1 - \xi\kappa_1.$$

An alternative method of getting a third order accurate formula is to look for a match by writing

$$I_b = \bar{\kappa}_0 f(\tilde{\mathbf{0}}) + \sum_{j=1}^{2} \kappa_1^j \Sigma_1(\lambda_1^j)$$

$$= \bar{\kappa}_0 A + \sum_{j=1}^{2} \kappa_1^j [\xi A + (\lambda_1^j)^2 C_2 + (\lambda_1^j)^3 D_3]$$

$$= A + \frac{p!}{(p+2)!} C_2 + \frac{2p!}{(p+3)!} D_3 + \cdots, \qquad (5.43)$$

with j in κ_1^j and λ_1^j a superscript label, not an exponent. Equating coefficients of A we get

$$\bar{\kappa}_0 = 1 - \xi \sum_{j=1}^{2} \kappa_1^j, \qquad (5.44)$$

while equating coefficients of C_2 and D_3, we obtain a system of two simultaneous equations for κ_1^j, $j = 1, 2$,

$$q_1 = w_1 + w_2,$$
$$q_2 = \lambda_1^1 w_1 + \lambda_1^2 w_2, \qquad (5.45)$$

where we have abbreviated

$$q_1 = \frac{p!}{(p+2)!}, \qquad q_2 = \frac{2p!}{(p+3)!},$$
$$w_j = \kappa_1^j (\lambda_1^j)^2, \qquad j = 1, 2. \qquad (5.46)$$

This set of equations (forming a $N = 2$ Vandermonde system) can be immediately solved to give

$$w_1 = \frac{\lambda_1^2 q_1 - q_2}{\lambda_1^2 - \lambda_1^1},$$
$$w_2 = \frac{\lambda_1^1 q_1 - q_2}{\lambda_1^1 - \lambda_1^2}, \qquad (5.47)$$

giving a second third order accurate formula for any non-degenerate λ_1^1 and λ_1^2 lying in the interval $(0, 1)$.

5.5.3 *Fourth order simplex formula*

This follows a different pattern from the odd order formulas, and details are given in Appendix G.

5.5.4 *Fifth order simplex formula*

We turn next to deriving a fifth order formula. Referring to Eq. (5.35), we see that to get a fifth order formula we have to use the sums of Eq. (5.36) to match the coefficients of A, C_2, D_3, E_4, E_{2+2}, F_5, and F_{3+2}. Since at most double (i.e., two-integer) partitions appear, we can still get the leading double partition terms from $\Sigma_2(\lambda, \lambda)$, but we must now impose a condition on λ to guarantee that E_{2+2} and F_{3+2} appear with coefficients in the correct ratio. From Eq. (5.35) we see that the ratio of the coefficient of F_{3+2} to that of E_{2+2} must be $2/(p+5)$, and from Eq. (5.36) together with Eq. (F.1) of Appendix F, with $\lambda = \sigma$ and $\mu = \kappa = 0$, we see that this is obtained with

$$\lambda = \frac{2}{p+5},\qquad(5.48)$$

which for any $p \geq 1$ obeys the condition $2\lambda < 1$. The overall coefficient of Σ_2 needed to fit E_{2+2} and F_{3+2} is easily seen to be

$$\kappa_2 = \frac{p!}{2!(p+4)!}\lambda^{-4} = \frac{(p+5)^4}{32p_{41}},\qquad(5.49)$$

where we have used the abbreviated notation, for $n > m$,

$$p_{nm} \equiv (p+m)(p+m+1)\cdots(p+n).\qquad(5.50)$$

Thus we have, again from Eq. (F.1),

$$\kappa_2\Sigma_2(\lambda, \lambda) = \frac{p!}{(p+4)!}E_{2+2} + 2\frac{p!}{(p+5)!}F_{3+2} + \kappa_2[\xi^2 A + \xi(2\lambda^2 C_2$$

$$+ 2\lambda^3 D_3 + 2\lambda^4 E_4 + 2\lambda^5 F_5)],\qquad(5.51)$$

with $\xi = p + 1$. Since there are four single partition terms, we look for an integration formula of the form

$$\kappa_2\Sigma_2(\lambda, \lambda) + \sum_{j=1}^{4} \kappa_1^j \Sigma_1(\lambda_1^j) + \kappa_0 A,\qquad(5.52)$$

which is to be equated to the sum of terms through fifth order in Eq. (5.35).

The equation for matching the coefficient of A can immediately be solved in terms of the coefficients κ_2 and κ_1^j, giving

$$\kappa_0 = 1 - R_0, \quad R_0 = \xi^2 \kappa_2 + \xi \sum_{j=1}^{4} \kappa_1^j. \tag{5.53}$$

The four equations for matching the coefficients of C_2, D_3, E_4, and F_5 give a $N = 4$ Vandermonde system that determines the four coefficients κ_1^j. In terms of the standard form for the order N Vandermonde system given in Eq. (5.22), the equations determining the coefficients κ_1^j are

$$y_j = \lambda_1^j, \quad w_j = \kappa_1^j (\lambda_1^j)^2,$$

$$q_1 = \frac{1}{p_{21}} - 2\xi\kappa_2\lambda^2,$$

$$q_2 = \frac{2}{p_{31}} - 2\xi\kappa_2\lambda^3, \tag{5.54}$$

$$q_3 = \frac{6}{p_{41}} - 2\xi\kappa_2\lambda^4,$$

$$q_4 = \frac{24}{p_{51}} - 2\xi\kappa_2\lambda^5.$$

Solving this system of linear equations, for any non-degenerate values of the parameters $0 < \lambda_1^j < 1$, gives the coefficients κ_1^j and completes specification of the integration formula.

5.5.5 *Seventh order simplex formula*

To get a seventh order formula, we use the sums of Eq. (5.36) to match the coefficients appearing in Eq. (5.35) through the term H_{3+2+2}. Since at most triple partitions appear, we can get the leading triple partition terms G_{2+2+2} and H_{3+2+2} from $\Sigma_3(\lambda, \lambda, \lambda)$ by imposing the condition

$$\lambda = \frac{2}{p+7}, \tag{5.55}$$

which guarantees that their coefficients are in the correct ratio, and which for any $p \geq 1$ obeys the condition $3\lambda < 1$. The overall coefficient of Σ_3 needed to fit G_{2+2+2} and H_{3+2+2} is

$$\kappa_3 = \frac{p!}{3!(p+6)!}\lambda^{-6} = \frac{(p+7)^6}{384 p_{61}}.$$ (5.56)

We now look for an integration formula of the form

$$\kappa_3 \Sigma_3(\lambda,\lambda,\lambda) + \kappa_2' \Sigma_2(2\lambda,\lambda) + \sum_{j=1}^{U} \kappa_2^j \Sigma_2(\lambda_2^j, \lambda_2^j) + \sum_{j=1}^{6} \kappa_1^j \Sigma_1(\lambda_1^j) + \kappa_0 A,$$ (5.57)

with $U \leq 6$ since there are six double partition terms to be matched. Equating coefficients of the double partition terms, we find that the equations for $G_{4+2} - G_{3+3}$ and $H_{5+2} - H_{4+3}$ are both automatically satisfied by taking

$$\kappa_2' = 3\kappa_3.$$ (5.58)

This leaves only the double partition terms E_{2+2}, F_{3+2}, G_{4+2}, and H_{5+2} to be matched, so we can take the upper limit in the Σ_2 summation as $U = 4$. The four coefficients κ_2^j are then determined by solving an $N = 4$ Vandermonde system with inhomogeneous terms $q2_j$, $j = 1, \ldots, 4$, again with $\xi = p + 1$,

$$y_j = \lambda_2^j, \qquad w_j = 2\kappa_2^j(\lambda_2^j)^4,$$

$$q2_1 = \frac{1}{p_{41}} - (6\xi + 24)\kappa_3\lambda^4,$$

$$q2_2 = \frac{2}{p_{51}} - (6\xi + 36)\kappa_3\lambda^5,$$ (5.59)

$$q2_3 = \frac{6}{p_{61}} - (6\xi + 60)\kappa_3\lambda^6,$$

$$q2_4 = \frac{24}{p_{71}} - (6\xi + 108)\kappa_3\lambda^7.$$

We next have to match the six single partition terms, using Σ_1 sums. To save function calls, we take four of the parameters λ_1^j to be equal to $2\lambda_2^j$, with the other two λ_1^j taken as new, independent parameters. Equating the coefficients of the single partition terms

C_2 through H_7 then gives a $N = 6$ Vandermonde system determining the coefficients κ_1^j, with inhomogeneous terms $q1_j$, $j = 1, \ldots, 6$,

$$y_j = \lambda_1^j, \quad w_j = \kappa_1^j(\lambda_1^j)^2,$$

$$q1_1 = \frac{1}{p_{21}} - 2\xi \sum_{j=1}^{4} \kappa_2^j(\lambda_2^j)^2 - (3\xi^2 + 15\xi)\lambda^2\kappa_3,$$

$$q1_2 = \frac{2}{p_{31}} - 2\xi \sum_{j=1}^{4} \kappa_2^j(\lambda_2^j)^3 - (3\xi^2 + 27\xi)\lambda^3\kappa_3,$$

$$q1_3 = \frac{6}{p_{41}} - \xi q2_1 - (3\xi^2 + 51\xi)\lambda^4\kappa_3, \qquad (5.60)$$

$$q1_4 = \frac{24}{p_{51}} - \xi q2_2 - (3\xi^2 + 99\xi)\lambda^5\kappa_3,$$

$$q1_5 = \frac{120}{p_{61}} - \xi q2_3 - (3\xi^2 + 195\xi)\lambda^6\kappa_3,$$

$$q1_6 = \frac{720}{p_{71}} - \xi q2_4 - (3\xi^2 + 387\xi)\lambda^7\kappa_3.$$

Note that in $q1_3, \ldots, q1_6$, the sums $2\sum_{j=1}^{4} \kappa_2^j(\lambda_2^j)^m$, $m = 4, \ldots, 7$ have been eliminated in terms of $q2_1, \ldots, q2_4$ by using the Vandermonde system of Eq. (5.59). Finally, matching the coefficient of A we get, using Eq. (5.58)

$$\kappa_0 = 1 - R_0, \quad R_0 = (\xi^3 + 3\xi^2)\kappa_3 + \xi^2 \sum_{j=1}^{4} \kappa_2^j + \xi \sum_{j=1}^{6} \kappa_1^j. \quad (5.61)$$

5.5.6 *Ninth order simplex formula*

A sketch of the ninth order case (but not full formulas, which can be read off from the computer program) is given in Appendix H. Truncation errors become significant for the ninth order formula in double precision $\big(\text{real}(8)\big)$; to exploit its capabilities we recommend use of quadruple precision $\big(\text{real}(16)\big)$.

5.5.7 *Leading term in higher order for simplexes*

We have not systematically pursued constructing integration formulas of orders higher than ninth, but this should be possible by the same method. One can, however, see what the pattern will be for the leading term in such formulas. An integration formula of order $2t + 1$ will have a leading term $\Sigma_t(\lambda, \ldots, \lambda)$, with t arguments λ. The only t-integer partition terms appearing in the continuation of Eq. (5.35) will be $2 + 2 + \cdots + 2$, containing t terms 2, and $3 + 2 + \cdots + 2$, with one 3 and $t - 1$ terms 2. Requiring these to have coefficients in the correct ratio restricts λ to be

$$\lambda = \frac{2}{p + 2t + 1},$$ (5.62)

and the leading term in the integration formula will be $\kappa_t \Sigma_t(\lambda, \ldots, \lambda)$, with κ_t given by

$$\kappa_t = \frac{p!}{t!(p + 2t)!\lambda^{2t}}.$$ (5.63)

Where non-leading terms give multiple equations of the same order, corresponding to inequivalent partitions of $2t + 1, 2t, \ldots$, one has to include terms proportional to $\Sigma_{t-1}(2\lambda, \lambda, \ldots, \lambda)$, $\Sigma_{t-2}(3\lambda, \lambda, \ldots, \lambda)$, and other such structures with asymmetric arguments summing to $t\lambda$, for the differences of these multiple equations to have consistent solutions. Once such multiplicities have been taken care of, the remaining independent equations will form a number of sets of Vandermonde equations.

5.6 Parameterized higher order integration formulas for axis-parallel hypercubes from moments

We turn in this section to the problem of deriving higher order integration formulas for axis-parallel hypercubes, in analogy with our treatment of the simplex case. Our formulas can be viewed as a generalization of those obtained by Lyness (1965) and McNamee and Stenger (1967). We consider an axis-parallel hypercube of half-side S, and denote by $\tilde{\mathbf{x}}$ coordinates referred to the centroid of the hypercube. Through ninth order, the expansion of a general function $f(\tilde{\mathbf{x}})$

is given as before by Eq. (5.25). Consider now the moment integrals

$$m(\nu) = \int_{\text{hypercube}} dx_1 \cdots dx_p \, \tilde{x}_1^{\nu_1} \cdots \tilde{x}_p^{\nu_p} . \tag{5.64}$$

Since the limits of integration for each axis are $-S$, S, the moment integral factorizes and can be immediately evaluated as

$$m(\nu) = \prod_{\ell=1}^{p} \frac{S^{\nu_\ell + 1}}{\nu_\ell + 1} [1 + (-1)^{\nu_\ell}]$$

$$= 0 \quad \text{any } \nu_\ell \text{ odd},$$

$$= \prod_{\ell=1}^{p} \frac{2S \, S^{\nu_\ell}}{\nu_\ell + 1} = V \prod_{\ell=1}^{p} \frac{S^{\nu_\ell}}{\nu_\ell + 1} \quad \text{all } \nu_\ell \text{ even}, \tag{5.65}$$

where $V = (2S)^p$ in the final line is the hypercube volume.

We now re-express this moment integral in terms of sums over the set of $2p$ hypercube points $\tilde{\mathbf{x}}_a$ given in Eq. (5.19), which will play a role in hypercube integration analogous to that played by simplex vertices in our treatment of simplex integration. In analogy with Eq. (5.31), we define the sum

$$S_{i_1 \cdots i_n} = \sum_{a=1}^{2p} \tilde{x}_{a i_1} \cdots \tilde{x}_{a i_n} . \tag{5.66}$$

Since $\tilde{x}_{ai} = S\delta_{ai}$ for $1 \leq a \leq p$ and $\tilde{x}_{ai} = -S\delta_{ai}$ for $p + 1 \leq a \leq 2p$, this sum vanishes unless n is even and all of the indices $i_1, \ldots, i_n$ are equal, in which case it is equal to $2S^n$. The tensors of Eq. (5.66) and their direct products form a complete basis on which we can expand moment integrals over the hypercube. We have carried out this calculation in two different ways. First, by matching the non-vanishing moment integrals through eighth order, we find

$$\frac{1}{V} \int_{\text{hypercube}} dx_1 \cdots dx_p \tilde{x}_{i_1} \tilde{x}_{i_2} = \frac{1}{6} S_{i_1 i_2},$$

$$\frac{1}{V} \int_{\text{hypercube}} dx_1 \cdots dx_p \tilde{x}_{i_1} \tilde{x}_{i_2} \tilde{x}_{i_3} \tilde{x}_{i_4}$$

$$= \frac{1}{36}(S_{i_1 i_2} S_{i_3 i_4} + S_{i_1 i_3} S_{i_2 i_4} + S_{i_1 i_4} S_{i_2 i_3}) - \frac{1}{15} S_{i_1 i_2 i_3 i_4}$$

$$= \frac{1}{36}(S_{i_1 i_2} S_{i_3 i_4} + 2\,\text{terms}) - \frac{1}{15} S_{i_1 i_2 i_3 i_4},$$

$$\frac{1}{V} \int_{\text{hypercube}} dx_1 \cdots dx_p \tilde{x}_{i_1} \tilde{x}_{i_2} \tilde{x}_{i_3} \tilde{x}_{i_4} \tilde{x}_{i_5} \tilde{x}_{i_6}$$

$$= \frac{1}{216}(S_{i_1 i_2} S_{i_3 i_4} S_{i_5 i_6} + 14\,\text{terms}) \tag{5.67}$$

$$- \frac{1}{90}(S_{i_1 i_2} S_{i_3 i_4 i_5 i_6} + 14\,\text{terms}) + \frac{8}{63} S_{i_1 i_2 i_3 i_4 i_5 i_6},$$

$$\frac{1}{V} \int_{\text{hypercube}} dx_1 \cdots dx_p \tilde{x}_{i_1} \tilde{x}_{i_2} \tilde{x}_{i_3} \tilde{x}_{i_4} \tilde{x}_{i_5} \tilde{x}_{i_6} \tilde{x}_{i_7} \tilde{x}_{i_8}$$

$$= \frac{1}{1296}(S_{i_1 i_2} S_{i_3 i_4} S_{i_5 i_6} S_{i_7 i_8} + 104\,\text{terms})$$

$$+ \frac{1}{225}(S_{i_1 i_2 i_3 i_4} S_{i_5 i_6 i_7 i_8} + 34\,\text{terms})$$

$$- \frac{1}{540}(S_{i_1 i_2} S_{i_3 i_4} S_{i_5 i_6 i_7 i_8} + 209\,\text{terms})$$

$$+ \frac{4}{189}(S_{i_1 i_2} S_{i_3 i_4 i_5 i_6 i_7 i_8} + 27\,\text{terms})$$

$$- \frac{8}{15} S_{i_1 i_2 i_3 i_4 i_5 i_6 i_7 i_8}.$$

Combining these formulas with Eq. (5.25), and using a similar condensed notation to that used in the simplex case (but with the contractions referring now to the sums over the hypercube of Eq. (5.66)), we have for the integral of a general function over the hypercube, through ninth order,

$$\frac{1}{V}\int_{\text{hypercube}} dx_1 \cdots dx_p f(\tilde{\mathbf{x}}) = A + \frac{1}{6}C_2 + \frac{1}{36}E_{2+2} - \frac{1}{15}E_4$$

$$+ \frac{1}{216}G_{2+2+2} - \frac{1}{90}G_{4+2} + \frac{8}{63}G_6$$

$$+ \frac{1}{1296}I_{2+2+2+2} + \frac{1}{225}I_{4+4} - \frac{1}{540}I_{4+2+2}$$

$$+ \frac{4}{189}I_{6+2} - \frac{8}{15}I_8 + \cdots . \qquad (5.68)$$

A second, and more general way, to obtain these results is to construct a generating function, analogous to that of Good and Gaskins used in the simplex case. This is done in Appendix C, with results in agreement with those that we have just obtained.

We now follow the procedure used before in the simplex case. We match the expansion of Eq. (5.68) to sums over the function f evaluated at points within the hypercube, this time constructing these sums using parameter multiples of the $2p$ points of Eq. (5.19), which are the centroids of the maximal boundary hypercubes. The formulas of Eq. (5.36) together with Eqs. (F.1) and (F.2) of Appendix F still apply, with sums that extended from 0 to p in the simplex case extending now from 1 to $2p$, and with $\xi = p+1$ in Eq. (F.1) replaced by $\xi = 2p$.

We proceed to set up integration formulas of first, third, fifth, seventh, and ninth order, for integrals over an axis-parallel hypercube. Since all odd order terms in the expansion of Eq. (5.25) integrate to zero by symmetry, to achieve this accuracy it suffices to perform a matching of the non-vanishing terms through zeroth, second, fourth, sixth, and eighth order, respectively. We will see that as a result of the absence of odd order terms, the higher order hypercube formulas are considerably simpler than their general simplex analogs.

5.6.1 *First and third order formulas*

We begin our derivation of odd order hypercube integration formulas with examples of first and third order accuracy, obtained by matching the first two terms in the expansion of Eq. (5.68),

$$I = A + \frac{1}{6}C_2 + \cdots .$$ (5.69)

Proceeding in direct analogy with the first order formulas of Eq. (5.37) in the simplex case, we get

$$\begin{aligned} I_a &= f(\tilde{\mathbf{0}}), \\ I_b &= \Sigma_1(\lambda)/\xi, \end{aligned}$$ (5.70)

with $\tilde{\mathbf{x}} = \tilde{\mathbf{0}}$ the centroid of the hypercube, and $\xi = 2p$. These are again analogs of the center-of-bin and trapezoidal methods for the one-dimensional case.

Similarly, in analogy with the second order accurate formula of Eq. (5.40) for the simplex, we get the third order accurate hypercube formula

$$I = \left(1 - \frac{\xi}{6\lambda^2}\right) f(\tilde{\mathbf{0}}) + \frac{1}{6\lambda^2}\Sigma_1(\lambda).$$ (5.71)

For any two non-degenerate values $\lambda = \lambda_a$ and $\lambda = \lambda_b$ in the interval $(0,1)$ this gives two different third order accurate estimates $I_{a,b}$ of the hypercube integral.

5.6.2 *Fifth order formula*

To get a fifth order formula, we have to use the sums of Eq. (5.36) to match the coefficients of A, C_2, E_{2+2}, and E_4 appearing on the first line of Eq. (5.68). Since there is now only one double partition term, E_{2+2}, we can extract it from $\Sigma_2(\lambda, \lambda)$ for any λ in the interval $(0, \frac{1}{2})$. So we look for a fifth order formula of the form

$$\kappa_2 \Sigma_2(\lambda, \lambda) + \sum_{j=1}^{2} \kappa_1^j \Sigma_1(\lambda_1^j) + \kappa_0 A.$$ (5.72)

Matching the coefficient of E_{2+2} gives

$$\kappa_2 = \frac{1}{72\lambda^4},$$ (5.73)

while matching the coefficient of A gives, with $\xi = 2p$,

$$\kappa_0 = 1 - R_0, \qquad R_0 = \xi^2 \kappa_2 + \xi \sum_{j=1}^{2} \kappa_1^j.$$ (5.74)

Matching the coefficients of C_2 and E_4 gives a $N = 2$ Vandermonde system $\big($c.f. Eq. (5.22)$\big)$ with

$$y_j = (\lambda_1^j)^2, \quad w_j = \kappa_1^j(\lambda_1^j)^2,$$

$$q_1 = \frac{1}{6} - \frac{\xi}{36\lambda^2},$$

$$q_2 = \frac{-1}{15} - \frac{\xi}{36}. \tag{5.75}$$

5.6.3   *Seventh order formula*

To get a seventh order formula, we have to match the coefficients appearing on the first two lines of Eq. (5.68). Since there is only one triple partition term, G_{2+2+2}, we can extract it from $\Sigma_3(\lambda, \lambda, \lambda)$ for any λ in the interval $(0, \frac{1}{3})$. We look for a seventh order formula of the form

$$\kappa_3 \Sigma_3(\lambda, \lambda, \lambda) + \sum_{j=1}^{2} \kappa_2^j \Sigma_2(\lambda_2^j, \lambda_2^j) + \sum_{j=1}^{3} \kappa_1^j \Sigma_1(\lambda_1^j) + \kappa_0 A, \tag{5.76}$$

with matching the coefficient of G_{2+2+2} requiring

$$\kappa_3 = \frac{1}{1296\lambda^6}. \tag{5.77}$$

To reduce the number of function calls, we take $\lambda_1^{j+1} = 2\lambda_2^j$, $j = 1, 2$, with only λ_1^1 an additional parameter. Matching the coefficients of the double partition terms E_{2+2} and G_{4+2} gives a $N = 2$ Vandermonde system with

$$y_j = (\lambda_2^j)^2, \quad w_j = 2\kappa_2^j(\lambda_2^j)^4,$$

$$q_1 = \frac{1}{36} - \frac{\xi}{216\lambda^2},$$

$$q_2 = -\frac{1}{90} - \frac{\xi}{216}, \tag{5.78}$$

again with $\xi = 2p$, while matching the coefficient of A gives

$$\kappa_0 = 1 - R_0, \quad R_0 = \xi^3 \kappa_3 + \xi^2 \sum_{j=1}^{2} \kappa_2^j + \xi \sum_{j=1}^{3} \kappa_1^j. \tag{5.79}$$

Matching coefficients of the single partition terms C_2, E_4, and G_6 gives the $N = 3$ Vandermonde system with

$$y_j = (\lambda_1^j)^2, \quad w_j = \kappa_1^j(\lambda_1^j)^2,$$

$$q_1 = \frac{1}{6} - \frac{\xi^2}{432\lambda^4} - 2\xi \sum_{j=1}^{2} \kappa_2^j(\lambda_2^j)^2,$$

$$q_2 = -\frac{1}{15} - \frac{\xi}{36} + \frac{\xi^2}{432\lambda^2},$$ (5.80)

$$q_3 = \frac{8}{63} + \frac{\xi}{90} + \frac{\xi^2}{432}.$$

5.6.4 *Ninth order formula*

The details of the ninth order hypercube formula are given in Appendix I. Truncation errors become significant for the ninth order formula in double precision $(\text{real}(8))$; to exploit its capabilities we recommend use of quadruple precision $(\text{real}(16))$.

5.6.5 *Leading term in higher order for hypercubes*

As in the simplex analysis, in the hypercube case we have not systematically pursued constructing integration formulas of orders higher than ninth, but this should be possible by the same method. Again, one can see what the pattern will be for the leading term in such formulas. An integration formula of order $2t + 1$ will have a leading term $\Sigma_t(\lambda, \ldots, \lambda)$, with t arguments λ. The only t-integer partition term appearing in the continuation of Eq. (5.68) will be $2+2+\cdots+2$, containing t terms 2. So λ can be taken to have any value in the interval $(0, 1/t)$, and the leading term in the integration formula will be $\kappa_t\Sigma_t(\lambda, \ldots, \lambda)$, with κ_t given by

$$\kappa_t = \frac{1}{t!6^t\lambda^{2t}}.$$ (5.81)

Where non-leading terms give multiple equations of the same order, corresponding to inequivalent partitions of $2t, \ldots$ one has to include terms proportional to $\Sigma_{t-2}(3\lambda, \lambda, \ldots, \lambda)$, and other such structures with asymmetric arguments summing to $t\lambda$, for the differences of these multiple equations to have consistent solutions. Once such multiplicities have been taken care of, the remaining independent equations will form a number of sets of Vandermonde equations.

5.6.6   *Redundant function calls in low dimensions p*

Let us now examine the number of function evaluations required by our general moment fitting formulas for hypercubes, when restricted to one dimension. The first order center-of-bin formula requires just the one function evaluation $f(\mathbf{0})$, and so is the same in all methods. The third order formula of Eq. (5.71) involves $f(\mathbf{0})$ and one $\Sigma_1(\lambda)$, and so uses three function values, in agreement with Eq. (4.15), whereas the Gaussian formula needs two function values. Turning to the fifth order formula of Eq. (5.72), calculation of $\Sigma_2(\lambda, \lambda)$ requires three function evaluations, calculation of each of the two $\Sigma_1(\lambda^j)$ requires two function evaluations, and evaluation of $f(\mathbf{0})$ requires one function evaluation, for a total of eight function evaluations. This is to be compared to the one-dimensional moment fitting formula of Eq. (4.16) which requires five function evaluations, and the Gaussian method, which requires three.

The reason that the fifth order integration formula for general p, when specialized to one dimension, requires more function evaluations than the moment fitting method of Eq. (4.16), is that whereas in two and higher dimensions W_4 and W_2^2 of Eq. (C.3) in Appendix C are linearly independent, in one dimension they are proportional to one another by virtue of the identity $t_1^4 = (t_1^2)^2$. Hence the term $\Sigma_2(\lambda, \lambda)$ in the general integration formula is not needed to get a match, and when this is dropped one has a formula identical in form to that of Eq. (4.16), requiring only three function calls. Turning to the higher order hypercube formulas, we see that the seventh order hypercube formula of Eq. (5.76) has redundant parameters and function calls for dimension $p < 3$, since in two dimensions W_6, W_2W_4 and W_2^3 of Eq. (C.3) are linearly dependent by virtue of the algebraic identity

$$0 = (t_1^2 + t_2^2)^3 - 3(t_1^2 + t_2^2)(t_1^4 + t_2^4) + 2(t_1^6 + t_2^6). \qquad (5.82)$$

Similarly, the ninth order hypercube formula of Eq. (I.1) of Appendix I has redundant parameters and function calls for dimension

$p < 4$, since in three dimensions W_8, W_4^2, W_2W_6, $W_2^2W_4$, and W_2^4 of Eq. (C.3) are linearly dependent by virtue of the identity

$$0 = (t_1^2 + t_2^2 + t_3^2)^4 - 6(t_1^8 + t_2^8 + t_3^8) + 3(t_1^4 + t_2^4 + t_3^4)^2$$
$$+ 8(t_1^2 + t_2^2 + t_3^2)(t_1^6 + t_2^6 + t_3^6) - 6(t_1^2 + t_2^2 + t_3^2)^2(t_1^4 + t_2^4 + t_3^4).$$

$$(5.83)$$

These results suggest the conjecture that the hypercube formula of order $2t+1$ will involve redundant parameters and function calls for dimension $p < t$, and we expect an analogous statement to apply for the simplex formulas derived above by the moment fitting method. This redundancy for small p is a consequence of the fact that the integration formulas that we have derived for simplexes and hypercubes are universal, in the sense that they involve the same number of parameters irrespective of the dimension p. As p increases, the number of sampling points increases, but the number of parameters, and the size of the Vandermonde systems needed to find coefficients, remains fixed.

5.7   Some details of the cube357 programs

The cube357 programs are a "hybrid" program that (i) allows use of a hyper-rectangular base region, (ii) permits subdivision of some subset $ip1 < ip = p$ of the axes, chosen according to various criteria, and (iii) allows a comparison of Monte Carlo with higher order evaluation of the subregions, to deal with functions that are not everywhere highly differentiable. We give a few details here of how the programs are constructed.

5.7.1   *Integration over hyper-rectangles*

Consider a hyper-rectangle with half-sides S_i, $i = 1, \ldots, p$, centered on the origin. The integral of a function $f(x_1, \ldots, x_p)$ over this region,

$$I = \int_{-S_1}^{S_1} dx_1 \cdots \int_{-S_p}^{S_p} dx_p f(x_1, \ldots, x_p), \qquad (5.84)$$

is mapped to an integral over the half-side 1 hypercube by writing $x_i = S_i y_i$, $i = 1, \ldots, p$, giving

$$I = \left(\prod_{i=1}^{p} S_i \right) \int_{-1}^{1} dy_1 \cdots \int_{-1}^{1} dy_p g(y_1, \ldots, y_p), \tag{5.85}$$

$$g(y_1, \ldots, y_p) = f(S_1 x_1, \ldots, S_p x_p) .$$

Once rewritten in this form, any of the higher order hypercube programs developed in Sec. 5.6 can be used to evaluate the subregion integral.

5.7.2 *Measure var(i) of variation along axis i*

The measure $var(i)$ of variation along axis i is formed by taking finite differences of the mapped function $g(\mathbf{y})$, using samplings given by the parameters set by the setparam subroutines. For example, given two sampling parameter values $\lambda < 1$, $\sigma < 1$, the second order and fourth order differences (corresponding to $idiff = 1$ and $idiff = 2$ respectively) for axis i of the function g of Eq. (5.85) are given by

$$idiff = 1 : \quad var(i) = |d(\lambda) + d(\sigma)|,$$

$$idiff = 2 : \quad var(i) = 24 \left| \frac{d(\lambda) - d(\sigma)}{\lambda^2 - \sigma^2} \right|,$$

$$d(\lambda) = \left[\frac{1}{2} \left(g(\lambda \hat{u}_i) + g(-\lambda \hat{u}_i) \right) - g(\mathbf{0}) \right] \Big/ \lambda^2, \tag{5.86}$$

$$d(\sigma) = \left[\frac{1}{2} \left(g(\sigma \hat{u}_i) + g(-\sigma \hat{u}_i) \right) - g(\mathbf{0}) \right] \Big/ \sigma^2,$$

with $\hat{u}_i$ the axis i unit vector of Eq. (5.18). In the programs, the same recipe is applied to hyper-rectangles with general center, and similar absolute values of finite differences are used to construct the sixth and eighth order differences corresponding to $idiff = 3$ and $idiff = 4$.

5.8   Some samples of code

The full PAMIR programs are accessible on the Internet, so we give here only a few selected samples, with added comment lines that are

not in the archived version. Comment lines precede the line or lines
of code to which they refer.

5.8.1   *Basic module*

We first give the basic adaptive module that appears in simplex-
subs579.for, which is repeated with modifications in all of the programs. The basic module appears once in all non-"r", "m" programs,
and twice in the "r" and "m" programs. For brevity, we have removed
here all lines pertaining to printing of output. As before, ip is the
spatial dimension p and $klim$ is $2^p - 1$. The limit on the number
of levels of subdivision is $llim$, eps is the error criterion, and i_init
selects whether to initialize to a standard or Kuhn simplex.

```
      subroutine adaptnew579s(ip,klim,llim,eps,i_init)
c adaptive integration over standard or Kuhn simplex
      implicit real(8)(a-h,m-z)
      implicit integer(i-l)
      integer(2) ivnew,ivertold
      integer(2) iscale
      dimension ivertold(1:ip,0:ip),vertold(1:ip,0:ip)
      dimension ivnew(1:ip,0:klim,0:ip)
      integer(2),allocatable :: ivstoreo(:,:,:), ivstorei(:,:,:)
      common/subdivision/isubdivision
      common/accuracy/iaccuracy
c ithin is the same as ithinlev of Sec. 2.4.1 and main programs
      common/thinning/ithin
      outinta=0.d0
      outintb=0.d0
      errorest=0.d0
      fcncallsadd=0.d0
      ivertold(1:ip,0:ip)=0
      ivnew(1:ip,0:klim,0:ip)=0
c iscale sets the maximum size of the integer(2) lattice
      iscale=16384
      rescale=16384.d0
      outa=0.d0
      outb=0.d0
      error=0.d0
      indmax=0
      fcncalls=0.d0
```

```
c  klimp  is  2**p
         klimp=klim+1
         aklimp=klimp
c  indx  counts  the  number  of  subregions  carried  to  next  level
         indx=1
c  initialization  for  standard  simplex  for  i_init=1
c  initialization  for  Kuhn  simplex  for  i_init=2
c  i  is  index  for  vector  component,  j  is  index
c  for  vertex;  ivertold  is  the  array  containing
c  the  base  region  vertices
         if ( i_init . eq . 1 )  then
               do  5  i =1,ip
                  ivertold ( i , i)=iscale
   5     continue
         else  if ( i_init . eq . 2 )  then
               do  6  i =1,ip
               do  6  j=i , ip
                  ivertold ( i , j)=iscale
   6     continue
         end  if
         vertold (1: ip ,0: ip)=ivertold (1: ip ,0: ip)/rescale
         volume =1.d0
c  recursion;  l  is  the  level  of  subdivision  number  referred
c  to  as  script  or  italic  l  in  the  text
         do  1000  l =1,llim
         if ( l . lt . llim )  then
c  indxmax  counts  the  maximum  number  of  subregions
c  that  must  be  allowed  for  in  the  output  array
c  ivstoreo  of  the  current  level ,  if  the  error
c  criterion  is  not  obeyed  for  any  subregion
                  indxmax=indx * klimp
                  allocate ( ivstoreo (1: ip ,0: ip ,1:indxmax))
                  ivstoreo (1: ip ,0: ip ,1:indxmax)=0
         end  if
         ind=0
c  ll  labels  the  particular  subregion  being  evaluated
         do  100  ll =1,indx
         if ( l . gt . 1 )  then
c  the  array  ivstorei  carried  forward  from  the
c  prior  level  is  used  to  create  the  array
c  ivertold  for  the  subregion  enumerated
c  by  ll  being  evaluated
```

```
                do 15  i=0,ip
                  ivertold(1:ip,i)=ivstorei(1:ip,i,ll)
   15      continue
c  the  integer  lattice  is  converted  to  real  numbers
                  vertold(1:ip,0:ip)=ivertold(1:ip,0:ip)/rescale
           end if
c  if  l  is  not  greater  than  ithin ,  program  subdivides  all
c  subregions ,  so  the  integration  routines  are  not  needed
           if(l.le.ithin.and.l.lt.llim) go to 125
c  the  appropriate  higher  order  routine
c  is  called ,  according  to  iaccuracy
           if(iaccuracy.eq.5) then
             call  simplexint5(ip,vertold,outinta,outintb,errorest,
   1      fcncallsadd)
           else if(iaccuracy.eq.7) then
             call  simplexint7(ip,vertold,outinta,outintb,errorest,
   1      fcncallsadd)
           else if(iaccuracy.eq.9) then
             call  simplexint9(ip,vertold,outinta,outintb,errorest,
   1      fcncallsadd)
           end if
           fcncalls=fcncalls+fcncallsadd
  125  continue
c  results  for  the  subregion  are  harvested  when
c  the  ''if''  is  satisfied
           if((l.eq.llim).or.(l.gt.ithin.and.errorest.lt.eps)) then
                  outa=outa+outinta*volume
                  outb=outb+outintb*volume
                  error=error+errorest*volume
c  the  subregion  is  subdivided  when  the  ''if''
c  is  not  satisfied
           else
                if(isubdivision.eq.1) then
                    call  symmetric579(ivnew,ivertold,ip,klim)
                else if(isubdivision.eq.2) then
                    call  recursive579(ivnew,ivertold,ip,klim)
                end if
                do 300 k=0,klim
                ind=ind+1
c  the  subdivided  regions  are  stacked  in  ivstoreo ,
c  the  output  stack  from  level  l
                  ivstoreo(1:ip,0:ip,ind)=ivnew(1:ip,k,0:ip)
```

```
 300  continue
         end if
 100  continue
         if(l.gt.1) deallocate(ivstorei)
         if(l.lt.llim) then
                 indx=ind
c the output stack from level l is transferred
c to ivstorei, which is the input stack for
c the next level; this is sized to accommodate
c indx subregions=final ind from level l,
c (which may be fewer than indxmax)
                 allocate(ivstorei(1:ip,0:ip,1:indx))
                 ivstorei(1:ip,0:ip,1:indx)=ivstoreo(1:ip,0:ip,1:indx)
                 deallocate(ivstoreo)
c the volume of the new subregions is calculated
                 volume=volume/aklimp
         end if
c indmax, not to be confused with indxmax,
c keeps track of the largest value of ind
         if(ind.gt.indmax) indmax=ind
         if(ind.eq.0) go to 5000
1000  continue
5000  continue
      return
      end
```

5.8.2 Simplex and hypercube rescaling for the "r" and "m" programs

Prior to the second stage of the "r" and "m" programs, each subregion of the integer(2) lattice that is carried forward in the stack *ivref* must be shifted and rescaled, so that it fills out an image of the original 16384 lattice used by the basic module, thus permitting further subdivision.

The first sample is for the simplex case.

```
c the following statements compute the shift needed
         iminn(1:ip)=16384
         do 9003 i=0,ip
         do 9994 k=1,ip
```

```
c fmin0 is an integer(2) external function which computes
c the minimum of its integer(2) arguments
        iminn(k)=fmin0(iminn(k),ivref(k,i,jjj))
 9994 continue
 9003 continue
        do 9004 i=0,ip
        imin(1:ip,i)=iminn(1:ip)
 9004 continue
c this do loop calculates the shifted
c and rescaled lattice
        do 6005 i=0,ip
        ivertold(1:ip,i)=(ivref(1:ip,i,jjj)-imin(1:ip,i))
   1 *2**(llim1+(1-isubdivision)*(2-i_init))
 6005 continue
c the corresponding real number values
c (undoing the shift and rescaling)
c are computed here; these are needed
c for evaluation of the integral in
c the next level
        vertold(1:ip,0:ip)=ivertold(1:ip,0:ip)/
   1 (rescale*2.d0**(llim1+(1-isubdivision)*(2-i_init)))
   2 +imin(1:ip,0:ip)/rescale
```

The next sample is the hypercube case.

```
c the half-side is rescaled
        ivertold(0)=ivref(0,jjj)*2**llim1
c shifting as accomplished by setting the
c centroid coordinates to the origin
        ivertold(1:ip)=0
c these lines calculate the corresponding
c real number values (undoing the shift
c and rescaling)
        vertold(0)=ivertold(0)/(rescale*2.d0**llim1)
        vertold(1:ip)=ivertold(1:ip)/(rescale*2.d0**llim1)
   1 +ivref(1:ip,jjj)/rescale
```

5.8.3   *Distributing residual subregions to processes in the "m" programs*

This sample shows the distribution of subregions to processes that appears in the "m" adaptive program in simplexsubs579m.for.

```
c  the following statements ''deal'' the subregions
c  out to the processes in round−robin fashion, so
c  that nearby subregions in the stack ivstorei that
c  is carried forward from the final level of the first
c  stage go to different processes; some simple modular
c  arithmetic is used to calculate indref, the size
c  of the stack carried to each process. The total
c  number of processes is iprocess; one serves as
c  an accumulation register for final answers from
c  the jprocess processes that receive subregions
          jprocess=iprocess−1
          ishare=indx/jprocess
          iresidue=indx−ishare*jprocess
          if(imy_rank.le.iresidue) then
                indref=ishare+1
          else if(imy_rank.gt.iresidue) then
                indref=ishare
          end if
          allocate(ivref(1:ip,0:ip,1:indref))
          do 7004 ll=1,ishare
          ivref(1:ip,0:ip,ll)=ivstorei(1:ip,0:ip,imy_rank+jprocess*(ll−1))
 7004 continue
          if(imy_rank.le.iresidue) then
                ivref(1:ip,0:ip,indref)=ivstorei(1:ip,0:ip,
    1     imy_rank+jprocess*ishare)
          end if
          deallocate(ivstorei)
c  the distribution of subregions is complete
c  loop over subregions
          do 8000 jjj=1,indref
c  this is the shift and rescaling given above for
c  the simplex case, now carried out within
c  each process
          iminn(1:ip)=16384
          do 9003 i=0,ip
          do 9994 k=1,ip
```

```
c  fmin0  is  an  integer(2)  external  function  which  computes
c  the  minimum  of  its  integer(2)  arguments
         iminn(k)=fmin0(iminn(k),ivref(k,i,jjj))
 9994  continue
 9003  continue
         do 9004  i=0,ip
         imin(1:ip,i)=iminn(1:ip)
 9004  continue
         do 6005  i=0,ip
         ivertold(1:ip,i)=(ivref(1:ip,i,jjj)-imin(1:ip,i))
   1  *2**(llim1+(1-isubdivision)*(2-i_init))
 6005  continue
         vertold(1:ip,0:ip)=ivertold(1:ip,0:ip)/
   1  (rescale*2.d0**(llim1+(1-isubdivision)*(2-i_init)))
   2  +imin(1:ip,0:ip)/rescale
```

5.9   Programming extensions and open questions

There are a number of possible extensions of the PAMIR programs
that could be pursued in the future. (1) The MPI version of the
programs could be rewritten to include redistribution of the process
workload after each level of the second stage. This would make the
reduction in running time when using thinning track more closely
with the reduction in the number of function calls. (2) The multi-
stage strategy could be extended to a third (or more) stages, by not
harvesting the subregions that fail to obey the thinning condition at
the end of the second stage, but instead writing them to a memory
device which is then read sequentially by a third stage, etc. (3) One
could implement the program flow diagrammed in Fig. 2.4(b) and
discussed in Sec. 2.7, which would avoid copying between two large
arrays. (4) One could build in an option of permuting the simplex
vertices at the start of the simplex programs, which gives a different
subdivision, and therefore a different evaluation of the integral for
use in estimating errors. (5) Finally, we remark that the same subdivision, thinning, and staging strategies that we have used will apply
with any integration formulas that give two different estimates of
the answer from each subregion, not just the parameterized moment

fitting formulas that we developed in Chapter 5.

There are also a number of mathematical questions that we have left open. (1) We found numerical evidence that symmetric subdivision of a standard simplex obeys the bound of Eq. (5.16) for reduction of side length, and that after ℓ symmetric (recursive) subdivisions, the resulting subsimplexes each fit within a hypercube of side $1/2^\ell$ $(1/2^{\ell-1})$. We do not have a proof of these conjectures, but have assumed them true in constructing the programs. (2) Given the regularities in the construction of parameterized fifth, seventh, and ninth order integration formulas for the simplex and hypercube cases, it would be of interest to try to find a general all-orders rule for these. (3) We have not addressed the question of analytic error estimates for the parameterized integration formulas. (4) We have not addressed in any systematic way the question of deciding which thinning function is optimal for a given choice of integrand. (5) It would be of interest to study the systematics, in the moment fitting method, of the tradeoff between the number of parameters that are fixed by appropriate conditions, and the number of function calls.

Appendix A

Test integrals

For verifying the higher order integration programs, and for checking the operation of the adaptive programs, it is essential to have test integrals with known answers. We summarize here useful formulas for various classes of integrals that can be used to test the PAMIR programs.

(1) Polynomial integrated over standard simplex
   We recall that the integral of a general function $f(x_1, \ldots, x_p)$ over a standard simplex is given by (c.f. Eq. (2.4))

$$\int_{\text{standard simplex}} dx_1 \cdots dx_p \, f(x_1, \ldots, x_p) = \int_0^1 dx_1 \int_0^{1-x_1} dx_2$$

$$\times \int_0^{1-x_1-x_2} dx_3 \cdots \int_0^{1-x_1-x_2-\cdots-x_{p-2}} dx_{p-1}$$

$$\times \int_0^{1-x_1-x_2-\cdots-x_{p-1}} dx_p f(x_1, \ldots, x_p).$$

$$(A.1)$$

When the function f is a product of power law factors, a useful formula is the multinomial beta function integral,

$$\int_{\text{standard simplex}} dx_1 \cdots dx_p \, (1 - x_1 - x_2 - \cdots - x_p)^{\alpha_0-1} x_1^{\alpha_1-1} \cdots x_p^{\alpha_p-1}$$

$$= \frac{\prod_{a=0}^p \Gamma(\alpha_a)}{\Gamma(\sum_{a=0}^p \alpha_a)}, \qquad (A.2)$$

with Γ the usual gamma function (see the Wikipedia article on the Dirichlet distribution). When $\alpha_a - 1 = \nu_a$, $a = 0, \ldots, p$ with ν_a a non-negative integer, this can be rewritten as

$$\int_{\text{standard simplex}} dx_1 \cdots dx_p \, (1 - x_1 - x_2 - \cdots - x_p)^{\nu_0} x_1^{\nu_1} \cdots x_p^{\nu_p}$$

$$= \frac{\prod_{a=0}^{p} \nu_a!}{(p + \sum_{a=0}^{p} \nu_a)!} \, . \tag{A.3}$$

The $\nu_0 = 0$ case of this formula is the formula given by Stroud (1971) (see also Grundmann and Möller (1978)) for the integral of a general monomial over the standard simplex. To get the volume normalized integral, one multiplies by $p!$.

(2) Polynomial integrated over unit and half-side 1 hypercubes
    For a unit hypercube, the corresponding formula (for $\nu_\ell > -1$) is

$$\int_0^1 dx_1 \cdots \int_0^1 dx_p \, x_1^{\nu_1} \cdots x_p^{\nu_p} = \prod_{\ell=1}^{p} \frac{1}{\nu_\ell + 1} \, , \tag{A.4}$$

which gives directly the volume normalized integral (and is valid for non-integral as well as integral ν_ℓ). For a half-side 1 hypercube the corresponding monomial integrals, for ν_ℓ a non-negative integer, are

$$\int_{-1}^1 dx_1 \cdots \int_{-1}^1 dx_p \, x_1^{\nu_1} \cdots x_p^{\nu_p} = 2^p \prod_{\ell=1}^{p} \frac{1}{\nu_\ell + 1} \quad \text{for all } \nu_\ell \text{ even,}$$

$$\text{and zero otherwise} \, , \tag{A.5}$$

which when divided by 2^p gives the volume normalized integral.

(3) Polynomial integrated over Kuhn simplex
    Here which variable gets which exponent matters, and we have (for $\nu_p > -1$, $\nu_p + \nu_{p-1} > -2$, etc.)

$$\int_0^1 dx_1 \int_0^{x_1} dx_2 \int_0^{x_2} dx_3 \cdots \int_0^{x_{p-1}} dx_p x_1^{\nu_1} x_2^{\nu_2} \cdots x_p^{\nu_p}$$

$$= \frac{1}{\nu_p + 1} \frac{1}{\nu_p + \nu_{p-1} + 2} \frac{1}{\nu_p + \nu_{p-1} + \nu_{p-2} + 3}$$

$$\times \cdots \times \frac{1}{\nu_p + \nu_{p-1} + \cdots + \nu_2 + \nu_1 + p} . \tag{A.6}$$

To get the volume normalized integral, one multiplies by $p!$. This formula is also valid for non-integral as well as integral ν_ℓ.

(4) Feynman–Schwinger integral

The Feynman–Schwinger formula for combining perturbation theory denominators,

$$\frac{1}{D_0 D_1 \cdots D_p} = p! \int_{\text{standard simplex}} dx_1 \cdots dx_p$$

$$\times \frac{1}{[(1 - x_1 - x_2 - \cdots - x_p) D_0 + x_1 D_1 + \cdots + x_p D_p]^{p+1}} , \tag{A.7}$$

can be proved inductively as follows. For $p = 1$, the Feynman–Schwinger formula reads

$$\frac{1}{D_0 D_1} = \int_0^1 dx_1 \frac{1}{[(1 - x_1) D_0 + x_1 D_1]^2} , \tag{A.8}$$

which is easily verified by carrying out the integral. Assume now that this formula holds for dimension p. For $p + 1$, the formula asserts that

$$\frac{1}{D_0 D_1 \cdots D_p D_{p+1}} = (p + 1)! \int_0^1 dx_1 \cdots \int_0^{1 - x_1 - x_2 - \cdots - x_{p-1}} dx_p$$

$$\times \int_0^{1 - x_1 - x_2 - \cdots - x_{p-1} - x_p} dx_{p+1}$$

$$\times \frac{1}{[(1 - x_1 - x_2 - \cdots - x_p - x_{p+1}) D_0 + x_1 D_1 + \cdots + x_p D_p + x_{p+1} D_{p+1}]^{p+2}} . \tag{A.9}$$

Carrying out the integral over x_{p+1}, we get

$$\frac{1}{D_0 D_1 \cdots D_p D_{p+1}} = p! \int_0^1 dx_1 \cdots \int_0^{1-x_1-x_2-\cdots-x_{p-1}} dx_p \frac{1}{D_{p+1} - D_0}$$

$$\times \left[\frac{1}{[(1 - \sum_{i=1}^p x_i) D_0 + x_1 D_1 + \cdots + x_p D_p]^{p+1}} \right.$$

$$\left. - \frac{1}{[(1 - \sum_{i=1}^p x_i) D_{p+1} + x_1 D_1 + \cdots + x_p D_p]^{p+1}} \right]. \qquad \text{(A.10)}$$

But applying the induction hypothesis for p dimensions, the right-hand side of this equation reduces to

$$\frac{1}{D_{p+1} - D_0} \left[\frac{1}{D_0 D_1 \cdots D_p} - \frac{1}{D_{p+1} D_1 \cdots D_p} \right] = \frac{1}{D_0 D_1 \cdots D_p D_{p+1}},$$

$$\text{(A.11)}$$

which is the result to be proved.

(5) Test integrals for the unit hypercube
    We consider a set of test integrals of the form

$$\int_{\text{unit hypercube}} dx_1 \cdots dx_p \, f(x_1, \ldots, x_p) = \int_0^1 dx_1 \cdots \int_0^1 dx_p f(x_1, \ldots, x_p),$$

$$\text{(A.12)}$$

with various choices of the function f as follows:

(i) The Tsuda function is based on the integral (with $a > 0$)

$$\int_0^1 dy \frac{a(a+1)}{(a+y)^2} = 1. \qquad \text{(A.13)}$$

In p dimensions this generalizes to the asymmetric test function (with $a_i > 0$)

$$f = \prod_{i=1}^p \left(\frac{a_i(a_i + 1)}{(a_i + x_i)^2} \right), \qquad \text{(A.14)}$$

which integrates to unity over the p-dimensional unit hypercube for any positive values of the a_i.

(ii) The product Lorentzian function is

$$f = \prod_{i=1}^p \frac{1}{\alpha_i^2 + (x_i - \beta_i)^2}, \qquad \text{(A.15)}$$

which for $0 \le \beta_i \le 1$ and $0 < \alpha_i$ has the integral over the p-dimensional hypercube

$$\int_{\text{unit hypercube}} dx_1 \cdots dx_p \, f$$

$$= \prod_{i=1}^{p} \frac{1}{\alpha_i} \left[\tan^{-1}\left(\frac{1-\beta_i}{\alpha_i}\right) + \tan^{-1}\left(\frac{\beta_i}{\alpha_i}\right) \right]. \quad (A.16)$$

For appropriate choices of the parameters β_i we can put the peak at a corner, an edge, a face, or a generic internal point of the hypercube.

(iii) The oscillatory function

$$f = \cos\left(\beta + \sum_{j=1}^{p} \alpha_j x_j\right) = \text{Re}[e^{i(\beta + \sum_{j=1}^{p} \alpha_j x_j)}], \quad (A.17)$$

when integrated over the unit hypercube, gives

$$\int_{\text{unit hypercube}} dx_1 \cdots dx_p \, f = \text{Re}\left[e^{i\beta} \prod_{j=1}^{p} \frac{e^{i\alpha_j} - 1}{i\alpha_j} \right]. \quad (A.18)$$

(iv) The C_0 function (which is continuous but not everywhere differentiable)

$$f = e^{-\sum_{i=1}^{p} |x_i - \beta_i|/a_i}, \quad (A.19)$$

is again a product of one-dimensional factors. For $0 \le \beta \le 1$ and $a > 0$, the one-dimensional integral

$$\int_0^1 dx e^{-|x-\beta|/a} = a[2 - e^{-\beta/a} - e^{-(1-\beta)/a}] \quad (A.20)$$

allows us to evaluate the integral of f over the p-dimensional hypercube,

$$\int_{\text{unit hypercube}} dx_1 \cdots dx_p \, f = \prod_{i=1}^{p} a_i[2 - e^{-\beta_i/a_i} - e^{-(1-\beta_i)/a_i}].$$

$$(A.21)$$

Appendix B

Derivation of the simplex generating function

We give here a proof of the simplex generating function formulas of Eqs. (5.27) and (5.28), using the standard simplex integral of Eq. (A.2), the simplex volume formula of Eq. (5.12), and the barycentric coordinate formulas of Eqs. (5.6) through (5.8). We start from

$$\int_{\text{simplex}} dx_1 \cdots dx_p \sum_{\nu_1 \cdots \nu_p = 0}^{\infty} \frac{\prod_{i=1}^{p} (\tilde{x}_i t_i)^{\nu_i}}{\prod_{i=1}^{p} \nu_i!} = \int_{\text{simplex}} dx_1 \cdots dx_p e^{\sum_{i=1}^{p} \tilde{x}_i t_i},$$

(B.1)

and substitute Eq. (5.6) on the right-hand side, giving

$$\int_{\text{simplex}} dx_1 \cdots dx_p e^{\sum_{a=0}^{p} \alpha_a \sum_{i=1}^{p} \tilde{x}_{ai} t_i}.$$

(B.2)

We now express the integral over the general simplex in terms of an integral over its barycentric coordinates α_a. Since $\sum_{a=0}^{p} \alpha_a = 1$, we can rewrite Eq. (5.8), by subtraction of $\mathbf{x}_0$ from both sides, as

$$\mathbf{x} - \mathbf{x}_0 = \sum_{a=0}^{p} (\mathbf{x}_a - \mathbf{x}_0) \alpha_a = \sum_{a=1}^{p} (\mathbf{x}_a - \mathbf{x}_0) \alpha_a.$$

(B.3)

From this we immediately find for the Jacobian

$$\left| \det \left(\frac{\partial x_1 \cdots \partial x_p}{\partial \alpha_1 \cdots \partial \alpha_p} \right) \right| = |\det(x_{ai} - x_{0i})| = Vp!,$$

(B.4)

with V the volume of the simplex. Since α_a, $a = 1, \ldots, p$ span a standard simplex, we have transformed the integral of Eq. (B.2) to the form

$$Vp! \int_{\text{standard simplex}} d\alpha_1 \cdots d\alpha_p e^{\sum_{a=0}^{p} \alpha_a \sum_{i=1}^{p} \tilde{x}_{ai} t_i}.$$

(B.5)

Expanding the exponential on the right in a power series, we have

$$Vp! \int_{\text{standard simplex}} d\alpha_1 \cdots d\alpha_p \sum_{\nu_1 \cdots \nu_p = 0}^{\infty} \frac{\prod_{a=0}^{p}(\alpha_a)^{\nu_a}(\sum_{i=1}^{p}\tilde{x}_{ai}t_i)^{\nu_a}}{\prod_{a=0}^{p}\nu_a!},$$

(B.6)

and then recalling that $\alpha_0 = 1 - \sum_{a=1}^{p}\alpha_a$, and using Eq. (A.3) to evaluate the integral over the standard simplex, we get

$$Vp! \sum_{\nu_1 \cdots \nu_p = 0}^{\infty} \frac{\prod_{a=0}^{p}(\sum_{i=1}^{p}\tilde{x}_{ai}t_i)^{\nu_a}}{(p + \sum_{a=0}^{p}\nu_a)!}.$$

(B.7)

Let us now define P_n as the projector on terms with a total of n powers of the parameters t_i, since this is the projector that extracts the nth order moments. Applying P_n to Eq. (B.7), the denominator is converted to $(p+n)!$, which can then be pulled outside the sum over the ν_a, permitting these sums to be evaluated as geometric series,

$$P_n Vp! \sum_{\nu_1 \cdots \nu_p = 0}^{\infty} \frac{\prod_{a=0}^{p}(\sum_{i=1}^{p}\tilde{x}_{ai}t_i)^{\nu_a}}{(p + \sum_{a=0}^{p}\nu_a)!}$$

$$= \frac{Vp!}{(p+n)!} P_n \sum_{\nu_1 \cdots \nu_p = 0}^{\infty} \prod_{a=0}^{p}\left(\sum_{i=1}^{p}\tilde{x}_{ai}t_i\right)^{\nu_a}$$

$$= \frac{Vp!}{(p+n)!} P_n \prod_{a=0}^{p}\left[1 - \left(\sum_{i=1}^{p}\tilde{x}_{ai}t_i\right)\right]^{-1}.$$

(B.8)

Finally, applying to each factor in the product over a the rearrangement

$$(1 - y)^{-1} = \exp[-\log(1 - y)] = \exp\left[\sum_{s=1}^{\infty}\frac{y^s}{s}\right],$$

(B.9)

we get

$$\frac{Vp!}{(p+n)!} P_n \exp\left[\sum_{s=2}^{\infty}\frac{\sum_{a=0}^{p}[\sum_{i=1}^{p}\tilde{x}_{ai}t_i]^s}{s}\right],$$

(B.10)

where we have used the fact that the $s = 1$ term in the sum vanishes because $\sum_{a=0}^{p}\tilde{x}_{ai} = 0$. Comparing Eq. (B.10) with P_n times the starting equation Eq. (B.1), we get Eqs. (5.27) and (5.28).

Appendix C

Derivation of the hypercube generating function

An alternative way to obtain the results of Eq. (5.68) is to construct a hypercube generating function, analogous to the simplex generating function of Appendix B. We start from the formula

$$V^{-1} \int_{-S}^{S} dx_1 \cdots \int_{-S}^{S} dx_p \, e^{t_1 x_1 + \cdots + t_p x_p} = \prod_{\ell=1}^{p} \frac{\sinh S t_\ell}{S t_\ell}, \qquad \text{(C.1)}$$

and recall the power series expansion for the logarithm of $\frac{\sinh x}{x}$,

$$\log\left(\frac{\sinh x}{x}\right) = \frac{1}{6} x^2 - \frac{1}{180} x^4 + \frac{1}{2835} x^6 - \frac{1}{37800} x^8 + \cdots$$

$$= \sum_{n=1}^{\infty} \frac{(-1)^{n+1} 2^{2n-1} B_{2n-1}}{n(2n)!} x^{2n}, \qquad \text{(C.2)}$$

with B_{2n-1} the Bernoulli numbers $B_1 = \frac{1}{6}$, $B_3 = \frac{1}{30}$, $B_5 = \frac{1}{42}$, $B_7 = \frac{1}{30}$, $B_9 = \frac{5}{66}, \ldots$. Defining, in analogy with the simplex case (with $\tilde{x}_{ai}$ now the components of the $2p$ vectors $\tilde{\mathbf{x}}_a$ of Eq. (5.19)),

$$W_u = \sum_{a=1}^{2p} \left(\sum_{i=1}^{p} \tilde{x}_{ai} t_i \right)^u$$

$$= 0, \quad u \text{ odd},$$

$$= 2 S^u \sum_{i=1}^{p} t_i^u, \quad u \text{ even}, \qquad \text{(C.3)}$$

we rewrite Eq. (C.1), using Eq. (C.2), as

$$V^{-1} \int_{-S}^{S} dx_1 \cdots \int_{-S}^{S} dx_p \, e^{t_1 x_1 + \cdots + t_p x_p} = e^{\sum_{n=1}^{\infty} K_n W_{2n}}, \qquad \text{(C.4)}$$

with

$$K_n = \frac{(-1)^{n+1}2^{2n-2}B_{2n-1}}{n(2n)!} .$$
(C.5)

Through eighth order, the right-hand side of Eq. (C.4) is

$e^{W_2/12 - W_4/360 + W_6/5670 - W_8/75600 + \cdots}$

$$\begin{aligned}
= \ & 1 \\
& + \frac{W_2}{12} \\
& - \frac{W_4}{360} + \frac{W_2^2}{288} \\
& + \frac{W_6}{5670} - \frac{W_2 W_4}{4320} + \frac{W_2^3}{10368} \\
& - \frac{W_8}{75600} + \frac{W_4^2}{259200} + \frac{W_2 W_6}{68040} - \frac{W_2^2 W_4}{103680} + \frac{W_2^4}{497664} \\
& + \cdots .
\end{aligned}$$
(C.6)

Applying the same rule as in the simplex case, of multiplying the coefficient of each term by a combinatoric factor $(s!)^t t!$ for each factor W_s^t, we recover the numerical coefficients in the expansions of Eqs. (5.67) and (5.68). This method can be readily extended to the higher order terms of these expansions.

Appendix D

Mappings between base regions

We summarize here mappings between base regions that are used in the programs or in derivations.

(1) *Hyper-rectangle onto half-side 1 hypercube*

As already noted in Eq. (2.8), the general hyper-rectangle is mapped onto a half-side 1 hypercube by the change of variable, for all i,

$$y_i = \frac{1}{2}(b_i + a_i) + x_i \frac{1}{2}(b_i - a_i) \,, \tag{D.1}$$

which maps the interval $-1 \leq x_i \leq 1$ onto the interval $a_i \leq y_i \leq b_i$. The corresponding Jacobian is

$$\left| \det \left(\frac{\partial y_1 \cdots \partial y_p}{\partial x_1 \cdots \partial x_p} \right) \right| = 2^{-p} \prod_{i=1}^{p} (b_i - a_i) \,. \tag{D.2}$$

(2) *General simplex to unit standard simplex*

Let $\mathbf{x}$ be an interior point of a general simplex with vertices $\mathbf{x}_a$, $a = 0, \ldots, p$. As already noted in our derivation of the simplex generating function, the expansion of $\mathbf{x}$ in barycentric coordinates α_a, $a = 0, \ldots, p$ given in Eq. (5.8),

$$\mathbf{x} = \sum_{a=0}^{p} \alpha_a \mathbf{x}_a \,,$$

$$1 = \sum_{a=0}^{p} \alpha_a, \quad \alpha_a \geq 0 \,, \tag{D.3}$$

maps the general simplex onto the standard simplex with coordinates $(\alpha_1, \ldots, \alpha_p)$. Using the formula for the volume V of the general simplex given in Eq. (5.12), we computed the Jacobian of this transformation in Eq. (B.4) to be

$$\left| \det \left(\frac{\partial x_1 \cdots \partial x_p}{\partial \alpha_1 \cdots \partial \alpha_p} \right) \right| = V p! \,. \tag{D.4}$$

If one wants to integrate over a general simplex, this transformation can be used to recast the integration into the standard simplex form calculated by the PAMIR simplex programs. The advantage of proceeding this way, rather than rewriting the programs to subdivide a general simplex base region, is that it allows the use of integer(2) arithmetic for storing the vertices of subdivided simplexes.

(3) *Stroud map from a unit standard simplex onto a unit hypercube*
In his classic book Stroud (1971) gave a map, which is widely used, from the standard simplex with coordinates $\mathbf{x} = (x_1, \ldots, x_p)$ to the side 1 hypercube with coordinates $\mathbf{y} = (y_1, \ldots, y_p)$,

$$\begin{aligned}
x_1 &= y_1 \,, \\
x_2 &= y_2(1 - y_1) \,, \\
x_3 &= y_3(1 - y_2)(1 - y_1) \,, \\
&\ldots\ldots\ldots\ldots\ldots \\
x_p &= y_p(1 - y_{p-1}) \cdots (1 - y_1) \,,
\end{aligned} \tag{D.5}$$

with inverse map

$$\begin{aligned}
y_1 &= x_1 \,, \\
y_2 &= x_2/(1 - x_1) \,, \\
y_3 &= x_3/(1 - x_1 - x_2) \,, \\
&\ldots\ldots\ldots\ldots\ldots \\
y_p &= x_p/(1 - x_1 - x_2 - \cdots - x_{p-1}) \,.
\end{aligned} \tag{D.6}$$

Comparing with Eq. (A.1), for $\mathbf{x}$ lying within a standard simplex the limits of integration for x_i are

$$\begin{aligned}
0 &\le x_1 \le 1 \,, \\
0 &\le x_i \le 1 - x_1 - \cdots - x_{i-1} \,, \quad i = 2, \ldots, p \,,
\end{aligned} \tag{D.7}$$

and so the corresponding limits of integration for y_i are

$$0 \le y_i \le 1, \quad i = 1, \ldots, p; \tag{D.8}$$

that is, **y** lies within a side 1 hypercube. From Eqs. (D.5) and (D.6) we find for the Jacobian of the Stroud map

$$\left| \det \left(\frac{\partial x_1 \cdots \partial x_p}{\partial y_1 \cdots \partial y_p} \right) \right| = (1 - y_1)^{p-1}(1 - y_2)^{p-2} \cdots (1 - y_{p-1}), \tag{D.9}$$

and for the Jacobian of the inverse Stroud map

$$\left| \det \left(\frac{\partial y_1 \cdots \partial y_p}{\partial x_1 \cdots \partial x_p} \right) \right| = \frac{1}{1 - x_1} \frac{1}{1 - x_1 - x_2} \cdots \frac{1}{1 - x_1 - x_2 - \cdots - x_{p-1}}. \tag{D.10}$$

(4) *Constant Jacobian map from a unit standard simplex onto a unit hypercube*

A constant Jacobian map from the unit standard simplex onto the unit (side 1) hypercube has been given by Genz and Cools (2003), drawing on methods in the book of Fang and Wang (1994). With **x** lying in the unit standard simplex, and **y** lying in the unit hypercube, the map is

$$x_p = 1 - y_p^{1/p},$$

$$x_{p-1} = y_p^{1/p}(1 - y_{p-1}^{1/(p-1)}),$$

$$x_{p-2} = y_p^{1/p} y_{p-1}^{1/(p-1)}(1 - y_{p-2}^{1/(p-2)})$$

$$\cdots\cdots\cdots\cdots\cdots \tag{D.11}$$

$$x_2 = y_p^{1/p} y_{p-1}^{1/(p-1)} \cdots y_3^{1/3}(1 - y_2^{1/2}),$$

$$x_1 = y_p^{1/p} y_{p-1}^{1/(p-1)} \cdots y_2^{1/2}(1 - y_1),$$

with inverse map

$$y_i = \left(\frac{S_i}{S_i + x_i} \right)^i, \qquad S_i = 1 - \sum_{j=i}^{p} x_j. \tag{D.12}$$

From Eqs. (D.11) and (D.12) we find for the constant Jacobian map

$$\left| \det \left(\frac{\partial x_1 \cdots \partial x_p}{\partial y_1 \cdots \partial y_p} \right) \right| = 1/p!, \tag{D.13}$$

and for the Jacobian of the inverse constant Jacobian map

$$\left| \det \left(\frac{\partial y_1 \cdots \partial y_p}{\partial x_1 \cdots \partial x_p} \right) \right| = p!\,, \tag{D.14}$$

with the sign of the determinant given by $(-1)^p$. Although a constant Jacobian map is appealing in principle, we found empirically that for simplex integration by integration over a unit hypercube, the Stroud map of Eq. (D.5) works much better in practice.

(5) *tiling of the unit hypercube by $p!$ unit Kuhn simplexes*

As discussed in Eq. (2.13) of Sec. 2.1.3, the unit (side 1) hypercube can be tiled by $p!$ unit Kuhn simplexes, within each of which the map has a constant Jacobian equal to 1. Recapitulating, the integral of a function f over the unit hypercube is equal to the integral of the symmetrized function computed from f, integrated over the unit Kuhn simplex,

$$\int_0^1 dx_1 \int_0^1 dx_2 \cdots \int_0^1 dx_{p-1} \int_0^1 dx_p f(x_1, \ldots, x_p)$$

$$= \int_{\text{unit Kuhn simplex}} dx_1 \cdots dx_p$$

$$\times \sum_{p! \text{ permutations } P} f(x_{P(1)}, \ldots, x_{P(p)})\,. \tag{D.15}$$

This mapping is useful for moderate values of p, but above $p = 7$ the number of terms in the symmetrization sum becomes too large for it to be an efficient way to integrate over hypercubes.

Appendix E

Rule for determining where a point lies with respect to a simplex

We prove here the rule, based on the expansion of Eqs. (5.6) through (5.8), for determining where the point $\mathbf{x}$ lies with respect to the simplex: (1) If all of the α_a are strictly positive, the point lies inside the boundaries of the simplex; (2) If a coefficient α_a is zero, the point lies on the boundary hyperplane opposite to the vertex $\mathbf{x}_a$, and if several of the α_a vanish, the point lies on the intersection of the corresponding boundary hyperplanes; (3) If any coefficient α_a is negative, the point lies outside the simplex.

To derive this rule, we observe that a point $\mathbf{x}$ lies within the simplex only if it lies on the same side of each boundary hyperplane of the simplex as the simplex vertex opposite that boundary. Let us focus on one particular vertex of the simplex, which we label $\mathbf{x}_p$, so that the other p vertices are $\mathbf{x}_0, \ldots, \mathbf{x}_{p-1}$. These p vertices span an affine hyperplane, which divides the p-dimensional space into two disjoint parts, and constitutes the simplex boundary hyperplane opposite the simplex vertex $\mathbf{x}_p$. A general parameterization of this hyperplane takes the form

$$\mathbf{x} = \mathbf{x}_0 + \sum_{a=1}^{p-1} \beta_a (\mathbf{x}_a - \mathbf{x}_0) , \tag{E.1}$$

that is, we take $\mathbf{x}_0$ as a fiducial point on the hyperplane and add arbitrary multiples of a complete basis of vectors $\mathbf{x}_a - \mathbf{x}_0$ in the hyperplane. Rewriting Eq. (E.1) as

$$\mathbf{x} = \sum_{a=0}^{p-1} \gamma_a \mathbf{x}_a , \tag{E.2}$$

with $\gamma_0 = 1 - \sum_{a=1}^{p-1} \beta_a$ and $\gamma_a = \beta_a$, $a \geq 1$, we see that the p coefficients γ_a obey the condition

$$\sum_{a=0}^{p-1} \gamma_a = 1 \,. \tag{E.3}$$

By virtue of this condition, we can also write the hyperplane parameterization of Eq. (E.2) in terms of coordinates with origin at the simplex centroid,

$$\tilde{\mathbf{x}} = \sum_{a=0}^{p-1} \gamma_a \tilde{\mathbf{x}}_a \,. \tag{E.4}$$

We now wish to determine whether a general point $\tilde{\mathbf{x}}$ lies on the same side of this hyperplane as the vertex $\tilde{\mathbf{x}}_p$, or lies on the hyperplane, or lies on the opposite side from $\tilde{\mathbf{x}}_p$, by using the expansion of Eqs. (5.6) and (5.7), which we rewrite in the form

$$\tilde{\mathbf{x}} = \sum_{a=0}^{p-1} \alpha_a \tilde{\mathbf{x}}_a + \frac{\alpha_p}{p} \sum_{a=0}^{p-1} \tilde{\mathbf{x}}_a$$
$$- \frac{\alpha_p}{p} \sum_{a=0}^{p-1} \tilde{\mathbf{x}}_a + \alpha_p \tilde{\mathbf{x}}_p \,. \tag{E.5}$$

The first line on the right-hand side of Eq. (E.5) has the form of the hyperplane parameterization of Eq. (E.4), since by construction the coefficients add up to unity, and so this part of the right-hand side is a point on the boundary hyperplane opposite the vertex $\tilde{\mathbf{x}}_p$. The second line on the right-hand side of Eq. (E.5) can be rewritten, by using Eq. (5.5), as

$$\alpha_p \frac{p+1}{p} \tilde{\mathbf{x}}_p \,. \tag{E.6}$$

To appreciate the significance of this, we note that the centroid of the p points defining the boundary hyperplane is

$$\tilde{\mathbf{x}}_{h;c} = \frac{1}{p} \sum_{a=0}^{p-1} \tilde{\mathbf{x}}_a = -\frac{1}{p} \tilde{\mathbf{x}}_p \,, \tag{E.7}$$

where we have again used Eq. (5.5). Therefore the vector from the centroid of the points defining the hyperplane to the vertex $\tilde{\mathbf{x}}_p$ is

$$\tilde{\mathbf{x}}_p - \tilde{\mathbf{x}}_{h;c} = \frac{p+1}{p}\tilde{\mathbf{x}}_p . \tag{E.8}$$

So Eq. (E.6) tells us that the point $\tilde{\mathbf{x}}$ is displaced from the hyperplane by a vector parallel to that of Eq. (E.8), with its length rescaled by the factor α_p. Therefore, if $\alpha_p > 0$, the point $\tilde{\mathbf{x}}$ lies on the same side of the boundary hyperplane as the opposite vertex $\tilde{\mathbf{x}}_p$. If $\alpha_p = 0$, the point $\tilde{\mathbf{x}}$ lies on the boundary hyperplane, and if $\alpha_p < 0$, the point $\tilde{\mathbf{x}}$ lies on the opposite side of the boundary hyperplane from the vertex $\tilde{\mathbf{x}}_p$. Applying this argument to all $p+1$ vertices in turn gives the rule stated above.

Appendix F

Expansion for Σ_4

We give here the expansion of Σ_4 in terms of $f(\tilde{\mathbf{0}}) = A$ and the contractions $C_2, \ldots, J_{3+2+2+2}$ appearing in Eq. (5.35). Abbreviating $\xi = p + 1$ for simplexes (and with $\xi = 2p$ for hypercubes), we have

$$
\begin{aligned}
\Sigma_4 = {} & \xi^4 A + \xi^3 (\lambda^2 + \sigma^2 + \mu^2 + \kappa^2) C_2 + \xi^3 (\lambda^3 + \sigma^3 + \mu^3 + \kappa^3) D_3 \\
& + \xi^3 (\lambda^4 + \sigma^4 + \mu^4 + \kappa^4) E_4 + 2\xi^2 (\lambda^2 \sigma^2 + \lambda^2 \mu^2 + \lambda^2 \kappa^2 \\
& + \sigma^2 \mu^2 + \sigma^2 \kappa^2 + \mu^2 \kappa^2) E_{2+2} + \xi^3 (\lambda^5 + \sigma^5 + \mu^5 + \kappa^5) F_5 \\
& + \xi^2 (\lambda^2 \sigma^3 + \sigma^2 \lambda^3 + \lambda^2 \mu^3 + \mu^2 \lambda^3 + \lambda^2 \kappa^3 + \kappa^2 \lambda^3 \\
& + \sigma^2 \mu^3 + \mu^2 \sigma^3 + \sigma^2 \kappa^3 + \kappa^2 \sigma^3 + \mu^2 \kappa^3 + \kappa^2 \mu^3) F_{3+2} \\
& + \xi^3 (\lambda^6 + \sigma^6 + \mu^6 + \kappa^6) G_6 + \xi^2 (\lambda^4 \sigma^2 + \lambda^2 \sigma^4 + \lambda^4 \mu^2 \\
& + \lambda^2 \mu^4 + \lambda^4 \kappa^2 + \lambda^2 \kappa^4 + \sigma^4 \mu^2 + \sigma^2 \mu^4 + \sigma^4 \kappa^2 \\
& + \sigma^2 \kappa^4 + \mu^4 \kappa^2 + \mu^2 \kappa^4) G_{4+2} + 6\xi (\lambda^2 \sigma^2 \mu^2 + \lambda^2 \sigma^2 \kappa^2 \\
& + \lambda^2 \mu^2 \kappa^2 + \sigma^2 \mu^2 \kappa^2) G_{2+2+2} + 2\xi^2 (\lambda^3 \sigma^3 + \lambda^3 \mu^3 \\
& + \lambda^3 \kappa^3 + \sigma^3 \mu^3 + \sigma^3 \kappa^3 + \mu^3 \kappa^3) G_{3+3} + \xi^3 (\lambda^7 + \sigma^7 + \mu^7 + \kappa^7) H_7 \\
& + \xi^2 (\lambda^5 \sigma^2 + \lambda^2 \sigma^5 + \lambda^5 \mu^2 + \lambda^2 \mu^5 + \lambda^5 \kappa^2 + \lambda^2 \kappa^5 + \sigma^5 \mu^2 \\
& + \sigma^2 \mu^5 + \sigma^5 \kappa^2 + \sigma^2 \kappa^5 + \mu^5 \kappa^2 + \mu^2 \kappa^5) H_{5+2} + 2\xi (\lambda^3 \sigma^2 \mu^2 + \lambda^3 \sigma^2 \kappa^2 \\
& + \lambda^3 \mu^2 \kappa^2 + \sigma^3 \lambda^2 \mu^2 + \sigma^3 \lambda^2 \kappa^2 + \sigma^3 \mu^2 \kappa^2 + \mu^3 \lambda^2 \sigma^2 \\
& + \mu^3 \lambda^2 \kappa^2 + \mu^3 \sigma^2 \kappa^2 + \kappa^3 \lambda^2 \sigma^2 + \kappa^3 \lambda^2 \mu^2 + \kappa^3 \sigma^2 \mu^2) H_{3+2+2} \\
& + \xi^2 (\lambda^4 \sigma^3 + \lambda^3 \sigma^4 + \lambda^4 \mu^3 + \lambda^3 \mu^4 + \lambda^4 \kappa^3 + \lambda^3 \kappa^4 + \sigma^4 \mu^3 + \sigma^3 \mu^4 \\
& + \sigma^4 \kappa^3 + \sigma^3 \kappa^4 + \mu^4 \kappa^3 + \mu^3 \kappa^4) H_{4+3}
\end{aligned}
$$

$$+ \xi^3(\lambda^8 + \sigma^8 + \mu^8 + \kappa^8)I_8$$
$$+ \xi^2(\lambda^6\sigma^2 + \lambda^2\sigma^6 + \lambda^6\mu^2 + \lambda^2\mu^6 + \lambda^6\kappa^2 + \lambda^2\kappa^6 + \sigma^6\mu^2 + \sigma^2\mu^6$$
$$+ \sigma^6\kappa^2 + \sigma^2\kappa^6 + \mu^6\kappa^2 + \mu^2\kappa^6)I_{6+2} + 2\xi(\lambda^4\sigma^2\mu^2 + \lambda^4\sigma^2\kappa^2$$
$$+ \lambda^4\mu^2\kappa^2 + \sigma^4\lambda^2\mu^2 + \sigma^4\lambda^2\kappa^2 + \sigma^4\mu^2\kappa^2 + \mu^4\lambda^2\sigma^2 + \mu^4\lambda^2\kappa^2$$
$$+ \mu^4\sigma^2\kappa^2 + \kappa^4\lambda^2\sigma^2 + \kappa^4\lambda^2\mu^2 + \kappa^4\sigma^2\mu^2)I_{4+2+2}$$
$$+ 24\lambda^2\sigma^2\mu^2\kappa^2 I_{2+2+2+2} + \xi^2(\lambda^5\sigma^3 + \lambda^3\sigma^5 + \lambda^5\mu^3 + \lambda^3\mu^5 + \lambda^5\kappa^3$$
$$+ \lambda^3\kappa^5 + \sigma^5\mu^3 + \sigma^3\mu^5 + \sigma^5\kappa^3 + \sigma^3\kappa^5 + \mu^5\kappa^3 + \mu^3\kappa^5)I_{5+3}$$
$$+ 2\xi(\lambda^2\sigma^3\mu^3 + \lambda^2\sigma^3\kappa^3 + \lambda^2\mu^3\kappa^3 + \sigma^2\lambda^3\mu^3 + \sigma^2\lambda^3\kappa^3 + \sigma^2\mu^3\kappa^3$$
$$+ \mu^2\lambda^3\sigma^3 + \mu^2\lambda^3\kappa^3 + \mu^2\sigma^3\kappa^3 + \kappa^2\lambda^3\sigma^3 + \kappa^2\lambda^3\mu^3 + \kappa^2\sigma^3\mu^3)I_{3+2+2}$$
$$+ 2\xi^2(\lambda^4\sigma^4 + \lambda^4\mu^4 + \lambda^4\kappa^4 + \sigma^4\mu^4 + \sigma^4\kappa^4 + \mu^4\kappa^4)I_{4+4}$$
$$+ \xi^3(\lambda^9 + \sigma^9 + \mu^9 + \kappa^9)J_9 + \xi^2(\lambda^7\sigma^2 + \lambda^2\sigma^7 + \lambda^7\mu^2 + \lambda^2\mu^7 + \lambda^7\kappa^2$$
$$+ \lambda^2\kappa^7 + \sigma^7\mu^2 + \sigma^2\mu^7 + \sigma^7\kappa^2 + \sigma^2\kappa^7 + \mu^7\kappa^2 + \mu^2\kappa^7)J_{7+2}$$
$$+ 2\xi(\lambda^5\sigma^2\mu^2 + \lambda^5\sigma^2\kappa^2 + \lambda^5\mu^2\kappa^2 + \sigma^5\lambda^2\mu^2 + \sigma^5\lambda^2\kappa^2 + \sigma^5\mu^2\kappa^2$$
$$+ \mu^5\lambda^2\sigma^2 + \mu^5\lambda^2\kappa^2 + \mu^5\sigma^2\kappa^2 + \kappa^5\lambda^2\sigma^2 + \kappa^5\lambda^2\mu^2$$
$$+ \kappa^5\sigma^2\mu^2)J_{5+2+2} + 6(\lambda^3\sigma^2\mu^2\kappa^2 + \lambda^2\sigma^3\mu^2\kappa^2 + \lambda^2\sigma^2\mu^3\kappa^2$$
$$+ \lambda^2\sigma^2\mu^2\kappa^3)J_{3+2+2+2} + \xi(\lambda^4\sigma^3\mu^2 + \lambda^4\sigma^2\mu^3 + \lambda^4\sigma^3\kappa^2 + \lambda^4\sigma^2\kappa^3$$
$$+ \lambda^4\mu^3\kappa^2 + \lambda^4\mu^2\kappa^3 + \sigma^4\lambda^3\mu^2 + \sigma^4\lambda^2\mu^3 + \sigma^4\lambda^3\kappa^2 + \sigma^4\lambda^2\kappa^3$$
$$+ \sigma^4\mu^3\kappa^2 + \sigma^4\mu^2\kappa^3 + \mu^4\lambda^3\sigma^2 + \mu^4\lambda^2\sigma^3 + \mu^4\lambda^3\kappa^2 + \mu^4\lambda^2\kappa^3$$
$$+ \mu^4\sigma^3\kappa^2 + \mu^4\sigma^2\kappa^3 + \kappa^4\lambda^3\sigma^2 + \kappa^4\lambda^2\sigma^3 + \kappa^4\lambda^3\mu^2 + \kappa^4\lambda^2\mu^3$$
$$+ \kappa^4\sigma^3\mu^2 + \kappa^4\sigma^2\mu^3)J_{4+3+2} + \xi^2(\lambda^6\sigma^3 + \lambda^3\sigma^6 + \lambda^6\mu^3 + \lambda^3\mu^6$$
$$+ \lambda^6\kappa^3 + \lambda^3\kappa^6 + \sigma^6\mu^3 + \sigma^3\mu^6 + \sigma^6\kappa^3 + \sigma^3\kappa^6 + \mu^6\kappa^3 + \mu^3\kappa^6)J_{6+3}$$
$$+ \xi^2(\lambda^5\sigma^4 + \lambda^4\sigma^5 + \lambda^5\mu^4 + \lambda^4\mu^5 + \lambda^5\kappa^4 + \lambda^4\kappa^5 + \sigma^5\mu^4 + \sigma^4\mu^5$$
$$+ \sigma^5\kappa^4 + \sigma^4\kappa^5 + \mu^5\kappa^4 + \mu^4\kappa^5)J_{5+4} + 6\xi(\lambda^3\sigma^3\mu^3 + \lambda^3\sigma^3\kappa^3$$
$$+ \lambda^3\kappa^3\mu^3 + \sigma^3\mu^3\kappa^3)J_{3+3+3} \,. \tag{F.1}$$

The following formulas are useful in evaluating the sums $\Sigma_{2,3,4}$ with repeated arguments, using a minimum number of function calls.

$$\Sigma_4(\lambda, \lambda, \lambda, \lambda) = 24 \sum_{a<b<c<d} f\big(\lambda(\mathbf{x}_a + \mathbf{x}_b + \mathbf{x}_c + \mathbf{x}_d)\big)$$

$$+ 12 \sum_a \sum_{b\neq a,\, c\neq a,\, b<c} f\big(2\lambda\mathbf{x}_a + \lambda(\mathbf{x}_b + \mathbf{x}_c)\big)$$

$$+ 6 \sum_a \sum_{b<a} f\big(2\lambda(\mathbf{x}_a + \mathbf{x}_b)\big) + 4 \sum_a \sum_{b\neq a} f(3\lambda\mathbf{x}_a + \lambda\mathbf{x}_b)$$

$$+ \sum_a f(4\lambda\mathbf{x}_a) \,,$$

$$\Sigma_3(\lambda, \lambda, \lambda) = 6 \sum_{a<b<c} f\big(\lambda(\mathbf{x}_a + \mathbf{x}_b + \mathbf{x}_c)\big) + 3 \sum_a \sum_{b\neq a} f(2\lambda\mathbf{x}_a + \lambda\mathbf{x}_b)$$

$$+ \sum_a f(3\lambda\mathbf{x}_a) \,,$$

$$\Sigma_2(\lambda, \lambda) = 2 \sum_a \sum_{b<a} f\big(\lambda(\mathbf{x}_a + \mathbf{x}_b)\big) + \sum_a f(2\lambda\mathbf{x}_a) \,,$$

$$\Sigma_3(2\lambda, \lambda, \lambda) = 2 \sum_a \sum_{b\neq a,\, c\neq a,\, b<c} f\big(2\lambda\mathbf{x}_a + \lambda(\mathbf{x}_b + \mathbf{x}_c)\big)$$

$$+ 2 \sum_a \sum_{b\neq a} f(3\lambda\mathbf{x}_a + \lambda\mathbf{x}_b) + 2 \sum_a \sum_{b<a} f\big(2\lambda(\mathbf{x}_a + \mathbf{x}_b)\big)$$

$$+ \sum_a f(4\lambda\mathbf{x}_a) \,,$$

$$\Sigma_2(3\lambda, \lambda) = \sum_a \sum_{b\neq a} f(3\lambda\mathbf{x}_a + \lambda\mathbf{x}_b) + \sum_a f(4\lambda\mathbf{x}_a) \,,$$

$$\Sigma_2(2\lambda, \lambda) = \sum_a \sum_{b\neq a} f(2\lambda\mathbf{x}_a + \lambda\mathbf{x}_b) + \sum_a f(3\lambda\mathbf{x}_a) \,. \qquad \text{(F.2)}$$

Appendix G

Fourth order simplex formula

In Sec. 5.5 we focused on odd-order simplex formulas; here we derive a fourth order formula, which follows a different pattern. Referring to Eq. (5.35), we see that to get a fourth order formula we have to use the sums of Eq. (5.36) to match the coefficients of A, C_2, D_3, E_4, and E_{2+2}. Since only the final one of these, E_{2+2}, involves a double (i.e., two-integer) partition of 4, we can use $\Sigma_2(\lambda, \lambda)$ to extract this, with any positive value of $\lambda \leq \frac{1}{2}$. Since the simplex subdivision algorithm uses the midpoints $\frac{1}{2}(\mathbf{x}_a + \mathbf{x}_b)$ as the vertices of the subdivided simplex, an efficient way to proceed in this case is to take $\lambda = \frac{1}{2}$ in $\Sigma_2(\lambda, \lambda)$, so that what is needed is the function value at the midpoints, and to compute these function values as part of the subdivision algorithm. This also yields the function values at the vertices of the subdivided simplex. We can get A from $f(\mathbf{x}_c)$, and we can fit C_2, D_3, and E_4 by evaluating $\Sigma_1(\lambda)$ with three distinct values of λ. One of these values can be taken as $\lambda = 1$, corresponding to the function values at the simplex vertices. The other two are free parameters, and by making two different choices for one of these, we get two different fourth order evaluations of the integral.

We worked out the fourth order program before proceeding systematically to the odd order cases, and so used a different notation from that of Eq. (5.36). Let us write f_c, f_v, and f_s for the sums of function values at the centroid, the vertices, and the side midpoints,

$$f_c = f(\tilde{\mathbf{x}}_c)\,,$$

$$f_v = \frac{1}{p+1} \sum_a f(\tilde{\mathbf{x}}_a)\,,$$

$$f_s = \frac{1}{(p+1)p} \sum_{a \neq b} f\left(\frac{1}{2}(\tilde{\mathbf{x}}_a + \tilde{\mathbf{x}}_b)\right)\,. \tag{G.1}$$

As in Eq. (5.50) of Sec. 5.5.4 we use, for $n > m$, the definition

$$p_{nm} \equiv (p+m)(p+m+1)\cdots(p+n)\,. \tag{G.2}$$

A simple calculation then shows that through fourth order terms, we have

$$\frac{1}{V} \int_{\text{simplex}} dx_1 \cdots dx_p f(\tilde{\mathbf{x}}) - \frac{8p}{p_{42}}\left(f_s + \frac{1}{p}f_v\right)$$

$$= k_0 A + k_2 C_2 + k_3 D_3 + k_4 E_4\,, \tag{G.3}$$

with coefficients given by

$$k_0 = 1 - \frac{8(p+1)}{p_{42}}\,,$$

$$k_2 = \frac{1}{p_{21}} - \frac{4}{p_{42}}\,,$$

$$k_3 = \frac{2}{p_{31}} - \frac{2}{p_{42}}\,, \tag{G.4}$$

$$k_4 = \frac{6}{p_{41}} - \frac{1}{p_{42}}\,.$$

Defining now

$$f_\lambda = \frac{1}{p+1} \sum_a f(\lambda \tilde{\mathbf{x}}_a)\,, \tag{G.5}$$

so that $f_1 = f_v$, we find that through fourth order,

$$E_4 = \frac{1}{\lambda_1 - \lambda_2}(t_1 - t_2)\,,$$

$$t_j = \frac{p+1}{1-\lambda_j}\left[f_v - f_c - \frac{1}{\lambda_j^2}(f_{\lambda_j} - f_c)\right], \quad j = 1, 2\,,$$

$$D_3 = \frac{1}{2} \sum_{j=1}^{2} [t_j - (\lambda_j + 1)E_4]\,,$$

$$C_2 = (p+1)(f_v - f_c) - D_3 - E_4\,.$$

When substituted into Eq. (G.3), this gives a fourth order formula for the integral, with a second evaluation of the integral obtained by replacing λ_2 by a third, distinct value λ_3.

Appendix H

Ninth order simplex formula

To get a ninth order formula, we use the sums of Eq. (5.36) to match the coefficients appearing in Eq. (5.35) through the final exhibited term $J_{3+2+2+2}$. Since at most quadruple partitions appear, we can get the leading quadruple partition terms $J_{3+2+2+2}$ and $I_{2+2+2+2}$ from $\Sigma_4(\lambda, \lambda, \lambda, \lambda)$ by imposing the condition

$$\lambda = \frac{2}{p+9}, \tag{H.1}$$

which guarantees that their coefficients are in the correct ratio, and which for any $p \geq 1$ obeys the condition $4\lambda < 1$. The overall coefficient of Σ_4 needed to fit $J_{3+2+2+2}$ and $I_{2+2+2+2}$ is

$$\kappa_4 = \frac{p!}{4!(p+8)!}\lambda^{-8} = \frac{(p+9)^8}{6144\,p_{81}}. \tag{H.2}$$

We now look (with benefit of hindsight) for an integration formula of the form

$$\kappa_4\Sigma_4(\lambda, \lambda, \lambda, \lambda) + \kappa_3'\Sigma_3(2\lambda, \lambda, \lambda) + \kappa_2''\Sigma_2(3\lambda, \lambda) + \sum_{j=1}^{4} \kappa_3^j\Sigma_3(\lambda_3^j, \lambda_3^j, \lambda_3^j)$$

$$+ \sum_{j=1}^{4} \kappa_2^j\Sigma_2(2\lambda_3^j, \lambda_3^j) + \sum_{j=1}^{6} \bar{\kappa}_2^j\Sigma_2(\lambda_2^j, \lambda_2^j) + \sum_{j=1}^{4} \kappa_1^j\Sigma_1(3\lambda_3^j)$$

$$+ \sum_{j=1}^{4} \bar{\kappa}_1^j\Sigma_1(2\lambda_3^j) + \kappa_0 A, \tag{H.3}$$

with four of the λ_2^j taken equal to the four λ_3^j, and the other two λ_2^j additional parameters. (Again, we reuse parameters wherever similar structures are involved in Eq. (F.2), so as to save function calls.)

We proceed to sketch the remaining calculation, without writing down the detailed form of the resulting Vandermonde equations (which can be read off from the programs, and is fairly complicated). We begin with the triple partition terms. The equations for $J_{5+2+2} - J_{4+3+2}$, $J_{4+3+2} - J_{3+3+3}$, and $I_{4+2+2} - I_{3+3+2}$ are all automatically satisfied by taking

$$\kappa_3' = 6\kappa_4 . \tag{H.4}$$

This leaves four independent matching conditions for G_{2+2+2}, H_{3+2+2}, I_{4+2+2}, and J_{5+2+2}, which lead to a $N = 4$ Vandermonde system determining the coefficients κ_3^j. We turn next to the double partition terms. We find that the equations for $I_{6+2} - 4.5I_{5+3} + 3.5I_{4+4}$ and $J_{7+2} - 3.5J_{6+3} + 2.5J_{5+4}$ are automatically satisfied by taking

$$\kappa_2'' = 8\kappa_4 . \tag{H.5}$$

The four equations for $G_{4+2} - G_{3+3}$, $H_{5+2} - H_{4+3}$, $I_{6+2} - I_{4+4}$, and $J_{7+2} + J_{6+3} - 2J_{5+4}$ then give a $N = 4$ Vandermonde system determining the coefficients κ_2^j. The remaining independent equations matching double partition terms, for E_{2+2}, F_{3+2}, G_{4+2}, H_{5+2}, I_{6+2}, and J_{7+2}, then give a $N = 6$ Vandermonde system determining the coefficients $\bar{\kappa}_2^j$.

Turning to the single partition terms, the eight equations obtained by matching coefficients for C_2, D_3, E_4, F_5, G_6, H_7, I_8, and J_9 give a $N = 8$ Vandermonde system determining simultaneously the four coefficients κ_1^j and the four coefficients $\bar{\kappa}_1^j$. Finally, equating coefficients of A gives (with $\xi = p + 1$)

$$\kappa_0 = 1 - R, \quad R = (\xi^4 + 6\xi^3 + 8\xi^2)\kappa_4 + \xi^3 \sum_{j=1}^{4} \kappa_3^j$$

$$+ \xi^2 \left(\sum_{j=1}^{4} \kappa_2^j + \sum_{j=1}^{6} \bar{\kappa}_2^j \right) + \xi \sum_{j=1}^{4} (\kappa_1^j + \bar{\kappa}_1^j) . \tag{H.6}$$

Appendix I

Ninth order hypercube formula

To get a ninth order formula, we have to match the coefficients of all three lines of Eq. (5.68). Since there is only one quadruple partition term $I_{2+2+2+2}$, we can extract it from $\Sigma_4(\lambda, \lambda, \lambda, \lambda)$ for any λ in the interval $(0, \frac{1}{4})$. We look for a ninth order formula of the form

$$\kappa_4 \Sigma_4(\lambda, \lambda, \lambda, \lambda) + \sum_{j=1}^{2} \kappa_3^j \Sigma_3(\lambda_3^j, \lambda_3^j, \lambda_3^j) + \kappa_2' \Sigma_2(3\lambda, \lambda)$$

$$+ \sum_{j=1}^{3} \bar{\kappa}_2^j \Sigma_2(\lambda_2^j, \lambda_2^j) + \sum_{j=1}^{4} \kappa_1^j \Sigma_1(\lambda_1^j) + \kappa_0 A, \quad \text{(I.1)}$$

with matching the coefficient of $I_{2+2+2+2}$ requiring

$$\kappa_4 = \frac{1}{31104\lambda^8} . \quad \text{(I.2)}$$

To reduce the number of function calls, we take $\lambda_2^j = \lambda_3^j$, $j = 1, 2$, and $\lambda_1^j = 3\lambda_2^j$, $j = 1, 2, 3$, with only λ_2^3 and λ_1^4 additional parameters. Matching the coefficients of the triple partition terms G_{2+2+2} and I_{4+2+2} gives a $N = 2$ Vandermonde system with

$$y_j = (\lambda_3^j)^2, \qquad w_j = 6\kappa_3^j (\lambda_3^j)^6 ,$$

$$q3_1 = \frac{1}{216} - \frac{\xi}{1296\lambda^2} ,$$

$$q3_2 = -\frac{1}{540} - \frac{\xi}{1296} , \quad \text{(I.3)}$$

where $\xi = 2p$. Taking the difference of the matching equations for I_{6+2} and I_{4+4} determines κ_2' to be

$$\kappa_2' = \frac{237}{8164800\lambda^8} = \frac{237}{262.5}\kappa_4 \,, \tag{I.4}$$

while matching the coefficient of A gives

$$\kappa_0 = 1 - R_0, \quad R_0 = \xi^4\kappa_4 + \xi^3 \sum_{j=1}^{2} \kappa_3^j + \xi^2 \left(\kappa_2' + \sum_{j=1}^{3} \bar{\kappa}_2^j \right) + \xi \sum_{j=1}^{4} \kappa_1^j \,. \tag{I.5}$$

Matching the remaining independent double partition terms E_{2+2}, G_{4+2}, and I_{4+4} gives a $N = 3$ Vandermonde system with

$$y_j = (\lambda_2^j)^2, \quad w_j = 2\bar{\kappa}_2^j(\lambda_2^j)^4 \,,$$

$$q2_1 = \frac{1}{36} - \kappa_2' 18\lambda^4 - \frac{\xi^2}{2592\lambda^4} - 6\xi \sum_{j=1}^{2} \kappa_3^j(\lambda_3^j)^4 \,,$$

$$q2_2 = -\frac{1}{90} - \kappa_2' 90\lambda^6 - \frac{\xi}{216} + \frac{\xi^2}{2592\lambda^2} \,, \tag{I.6}$$

$$q2_3 = \frac{1}{225} - \kappa_2' 162\lambda^8 + \frac{\xi}{540} + \frac{\xi^2}{2592} \,.$$

Finally, matching coefficients of the single partition terms C_2, E_4, G_6, and I_8 gives a $N = 4$ Vandermonde system with

$$y_j = (\lambda_1^j)^2, \quad w_j = \kappa_1^j(\lambda_1^j)^2 \,,$$

$$q1_1 = \frac{1}{6} - \frac{\xi^3}{7776\lambda^6} - \kappa_2' 10\xi\lambda^2 - 3\xi^2 \sum_{j=1}^{2} \kappa_3^j(\lambda_3^j)^2 - 2\xi \sum_{j=1}^{3} \bar{\kappa}_2^j(\lambda_2^j)^2 \,,$$

$$q1_2 = -\frac{1}{15} - \frac{\xi^3}{7776\lambda^4} - \kappa_2' 82\xi\lambda^4 - 3\xi^2 \sum_{j=1}^{2} \kappa_3^j(\lambda_3^j)^4 - \xi\, q2_1 \,, \tag{I.7}$$

$$q1_3 = \frac{8}{63} - \frac{\xi^3}{7776\lambda^2} - \kappa_2' 730\xi\lambda^6 - \frac{1}{2}\xi^2\, q3_1 - \xi\, q2_2 \,,$$

$$q1_4 = -\frac{8}{15} - \frac{\xi^3}{7776} - \kappa_2' 6562\xi\lambda^8 - \frac{1}{2}\xi^2\, q3_2 - \xi\, q2_3 \,.$$

Bibliography

These references make no pretense to completeness; they list what I have found useful in constructing the algorithms discussed in this book. For research groups with a continuing program, I have listed only later publications that contain earlier references. Numbers in square brackets following each reference list the pages where that reference is cited.

Bauberger, S. and Böhm, M. (1995). Simple one-dimensional integral representations for two-loop self energies: the master diagram. *Nucl. Phys. B* **445**, 25–46. [60, 63, 66, 67]

Berntsen, J. and Espelid, T. O. (1991). Error estimation in automatic quadrature routines. *ACM Trans. Math. Software* **17**, 233–252. [112, 113]

Broadhurst, D. J. (1990). The master two-loop diagram with masses. *Z. Phys. C — Particles and Fields* **47**, 115–124. [60]

Cools, R. and Haegemans, A. (2003). Algorithm 824: CUBPACK: A package for automatic cubature; framework description. *ACM Trans. Math. Software* **29**, 287–296. [114, 115]

Edelsbrunner, H. and Grayson, D. R. (2000). Edgewise subdivision of a simplex. *Discrete Comput. Geom.* **24**, 707–719. [125]

Ezure, H. (2008). Attenuation function of radiation from disk source with direction in cosine distribution. *J. Nucl. Sci. and Technology* **45**, 773–783. [95]

Fang, K. T. and Wang, Y. (1994). *Number-theoretic Methods in Statistics*. (London: Chapman & Hall). [179]

Fujimoto, J. et al. (1992). Numerical approach to loop integrals, in *New Computing Techniques in Physics Research II*, D. Perret-Gallix, ed. (Singapore: World Scientific), pp. 625–630. [60]

Genz, A. (1986). Fully symmetric interpolatory rules for multiple integrals. *SIAM J. Numer. Anal.* **23**, 1273–1283. [110, 111]

Genz, A. and Cools, R. (2003). An adaptive numerical cubature algorithm for simplices. *ACM Trans. Math. Software* **29**, 297–308. [51, 110, 111, 114, 179]

Good, I. J. and Gaskins, R. A. (1969). Centroid method of integration. *Nature* **222**, 697–698. [137, 138, 152]

Good, I. J. and Gaskins, R. A. (1971). The centroid method of numerical integration. *Numer. Math.* **16**, 343–359. [137, 138, 152]

Grundmann, A. and Möller, N. M. (1978). Invariant integration formulas for the *n*-simplex by combinatorial methods. *SIAM J. Numer. Anal.* **15**, 282–290. [109, 142, 168]

Hahn, T. (2005). CUBA — a library for multi-dimensional numerical integration. arXiv:hep-ph/0404043. [114]

Heinen, J. A. and Niederjohn, R. J. (1997). Comments on "Inversion of the VanderMonde Matrix". *IEEE Signal Process. Lett.* **4**, 115. [135]

Kahaner, D. K. and Wells, M. B. (1979). An experimental algorithm for *N*-dimensional adaptive quadrature. *ACM Trans. Math. Software* **5**, 86–96. [115]

Kaprzyk, S. and Mijnarends, P. E. (1986). A simple linear analytic method for Brillouin zone integration of spectral functions in the complex energy region. *J. Phys. C: Solid State Physics* **19**, 1283–1292. [52]

Katsura, S. et al. (1971). Lattice Green's function. Introduction. *J. Math. Phys.* **12**, 892–895. [52]

Knoble, H. D. (1995). Website download of BestLex subroutine. [2, 6, 49]

Kreimer, D. (1991). The master two-loop two-point function. The general case. *Phys. Lett. B* **273**, 277–281. [60]

Krommer, A. R. and Ueberhuber, C. W. (1994). *Numerical Integration on Advanced Computer Systems*, Lecture Notes in Computer Science 848. (Berlin: Springer-Verlag), p. 183. [101]

Krommer, A. R. and Ueberhuber, C. W. (1998). *Computational Integration*. (Philadelphia: SIAM). [40, 101, 109, 110, 112, 113, 115]

Kuhn, H. W. (1960). Some combinatorial lemmas in topology. *IBM J. Res. and Dev.* **4**, 518–524. [9]

Kurihara, Y. and Kaneko, T. (2006). Numerical contour integration for loop integrals. *Computer Phys. Comm.* **174**, 530–539. [60]

Lepage, G. P. (1978). A new algorithm for adaptive multidimensional integration. *J. Comp. Phys.* **27**, 192–203. [51, 58, 114]

Lyness, J. N. (1965). Symmetric integration rules for hypercubes II. Rule projection and rule extension. *Math. Comp.* **19**, 394–407. [109, 142, 149]

Lyness, J. N. and Kaganove, J. J. (1976). Comments on the nature of automatic quadrature routines. *ACM Trans. Math. Software* **2**, 65–81. [112]

Malcolm, M. A. and Simpson, R. B. (1975). Local versus global strategies for adaptive quadrature. *ACM Trans. Math. Software* **1**, 129–146. [115]

McKeeman, W. M. (1962). Algorithm 145: Adaptive numerical integration by Simpson's rule. *Commun. ACM* **5**, 604. [103]

McNamee, J. and Stenger, F. (1967). Construction of fully symmetric numerical integration formulas. *Num. Math.* **10**, 327–344. [149]

Methfessel, M. S., Boon, M. H., and Mueller, F. M. (1983). Analytic-quadratic method of calculating the density of states. *J. Phys. C: Solid State Physics* **16**, L949–L954. [52]

Moore, D. (1992). Subdividing simplices, in *Graphics Gems III*, D. Kirk ed. (Boston: Academic Press), pp. 244–249 and pp. 534–535. See also Moore, D.

(1992), Simplicial mesh generation with applications, Cornell University dissertation. [115, 120, 121, 125, 128, 131]

Morita, T. and Horiguchi, T. (1971a). Lattice Green's functions for the cubic lattices in terms of the complete elliptic integral. *J. Math. Phys.* **12**, 981–986. [52]

Morita, T. and Horiguchi, T. (1971b). Calculation of the lattice Green's function for the bcc, fcc, and rectangular lattices. *J. Math. Phys.* **12**, 986–992. [52]

Neagoe, V.-E. (1996). Inversion of the Van der Monde matrix. *IEEE Signal Process. Lett.* **3**, 119–120. [135]

Osborne, M. R. (2001). *Simplicial Algorithms for Minimizing Polyhedral Functions* (Cambridge: Cambridge U. Press), p. 5. [118]

Passarino, G. and Uccirati, S. (2002). Algebraic-numerical evaluation of Feynman diagrams: two-loop self-energies. *Nucl. Phys. B* **629**, 97–187. [60]

Plaza, A. (2007). The eight-tetrahedra longest-edge partition and Kuhn triangulations. *Comp. & Math. with Applications* **54**, 426–433, Fig. 1. [12]

Pontryagin, L. S. (1952). *Foundations of Combinatorial Topology*, English translation. (Rochester: Graylock Press), pp. 10–12. [120]

Press, W. H., Teukolsky, S. A., Vetterling, W. T., and Flannery, B. P. (1992). *Numerical Recipes in Fortran 77*, 2nd Edition. (Cambridge: Cambridge University Press), pp. 82–85 and pp. 309–320. [51, 59, 136]

Schürer, R. (2008). HIntLib Manual, mint. sbg. ac. at/HIntLib/manual. pdf. [115]

Silvester, P. (1970). Symmetric quadrature formulae for simplexes. *Math. Comput.* **24**, 95–100. [110]

Stroud, A. H. (1971). *Approximate Calculation of Multiple Integrals.* (Englewood Cliffs: Prentice-Hall). [15, 21, 109, 142, 168, 178, 179, 180]

Wikipedia articles on: *Adaptive Simpson's Method, Barycentric Coordinates, Dirichlet Distribution, Gaussian Quadrature, Hypercube, Polynomial Long Division, Simplex.* [13, 69, 120, 168]

Yuasa, F. et al. (2008). Numerical evaluation of feynman integrals by a direct computation method. *XII International Workshop on Advanced Computing and Analysis Techniques in Physics Research*, Erice, Sicily, Italy, Nov. 3–7, and arXiv:0904.2823. [60, 63, 116]

Index